MW00474572

Raggedy Ann and More

(1)

Raggedy Ann, P. F. Volland Company, 1918.

Raggedy Ann and More

Johnny Gruelle's Dolls and Merchandise

PATRICIA HALL

Especially for Janie -
An extraordinary
doll artist!
Very Best Wishes!
Patricia Hall
10/21/01

Pelican Publishing Company
Gretna 2000

Library of Congress Cataloging-in-Publication Data

Hall, Patricia, 1948-
Raggedy Ann and more : Johnny Gruelle's dolls and merchandise / Patricia Hall.
cm.
Includes biographical references and index.
ISBN 1-56554-102-2 (hc : alk. paper)
Raggedy Ann and Andy dolls—Collectibles—Catalogs.
Gruelle, Johnny. 1880-1938—Catalogs. I. Title.
NK4894.3.R34H36 1998 98-25073
688.7'221'0973075—dc21 CIP

All items pictured in this book are collectibles, including dolls, toys, books, novelties, photographs, artwork, clippings, and ephemera. Values are relative and can vary depending on age, condition, rarity, and provenance, as well as the general condition of the collectibles market.

Published by Pelican Publishing Company, Inc.
1000 Burmaster, Gretna, Louisiana 70053
Manufactured in China

For Barry and David and Russell —
My shining stars

Contents

Preface and Acknowledgments

IT WAS EARLY SPRING 1969, and I was giddy with unexpected good news. My modest collection of books by Johnny Gruelle had placed second in a student book competition sponsored by the University of California at Santa Cruz. My accompanying essay examined Gruelle's artistry and writing style and carefully discussed each book in my collection. However, about Gruelle I could say very little, except that he had created Raggedy Ann and Andy. For after much searching for reference material, I had found only a handful of small articles, which focused much more on his famed rag dolls than on the artist and illustrator who had created them.

During the next several decades I collected more of Johnny Gruelle's books, picking up any Raggedy dolls and toys I found along the way. I also continued searching for information about Gruelle, baffled that though his rag dolls were well-known, he was not. In fact, most people drew a blank when his name was mentioned. Despite unusual talent and abundant output, Johnny Gruelle's life and work had, for some odd reason, gone unchronicled. This was enough to call me to the task.

In 1988, I met with the Gruelle family and began researching and writing in earnest. The result was an illustrated biography, *Johnny Gruelle, Creator of Raggedy Ann and Andy* (Pelican Publishing Company, 1993). My work on Gruelle's life story introduced me to the Johnny Gruelle *oeuvre*—a remarkable and bounteous one that included not only books, newspaper cartoons, comics, and magazine stories, but also dolls and toys and derivative products of every imaginable variety. Even more intriguing was how these myriad creations had become interwoven in a fascinating, at times perilous, merchandising history. This account—of how Gruelle's creations came to be and what happened to them once they found their way to the marketplace—became the focus of *Raggedy Ann and More*.

This book is dedicated to collectors, those devoted pilgrims who thrill to the sight of a tousled-

The author at work, by cartoonist Jerry Marcus. (Reprinted by permission of Jerry Marcus)

Johnny Gruelle with young admirers, 1930s. (Cox-Scheffey Collection)

Original Christmas card by Johnny Gruelle. (Anne laDue Hartman Kerr)

yarn head peeking out of a garage-sale box and whose pulses race upon spotting the familiar Gruelle book cover atop a shop shelf. However, it is also written for the librarians, scholars, and history buffs who want to know more about the range and substance of Johnny Gruelle's works and their significance in the larger arena of twentieth-century popular culture.

Documented here are the dolls, toys, and other merchandise created or directly inspired by Johnny Gruelle. His most popular books have also been included, since many of the published works provided essential foundations for (or were planned tie-ins to) related dolls and merchandise. For ease in referencing, much of the information has been organized in topical sidebars, charts, and tables. (Note: More extensive bibliographic information will be the focus of a forthcoming Pelican book detailing Gruelle's published works.)

Standing at the center of the Gruelle merchandising story are, of course, his best-known characters, the inimitable (though often imitated) Raggedy Ann and Andy. However, *Raggedy Ann and More* reaches beyond the Raggedys' popularity to explore Gruelle's other works as well—creations that, though not as well known as his rag dolls, are vital to the story.

And quite a story it is. Though marred by tragedy and disappointment and beset with shortfalls and setbacks, the Gruelle merchandising saga is one of bright ideas, business optimism, and handcrafted ideals. It is also a story of seized (as well as missed) opportunities, serendipitous (as well as deliberate) marketing, and perseverance in the face of great social and political change.

As with any history that was not adequately documented as it occurred, an accounting of Gruelle's creations must acknowledge some very real gaps in the historical record. Complicating this are the many facts that have become intermeshed with legends, some at the hand of the master yarn-spinner himself, Johnny Gruelle.

In the minds of overly zealous journalists and well-meaning collectors, what really happened has become further obscured by more glamorous speculations, in some cases repeated so many times that they have come to be part of the accepted body of knowledge. *Raggedy Ann and More* attempts to clarify the facts, relying on many previously undisclosed and unavailable primary sources (including manufacturing records, contracts, correspondence, and financial documents) to set the record straight.

Raggedy Ann and More chronicles the Gruelle merchandising story through the early 1960s, the time when the family-spawned Johnny Gruelle Company was sold to Bobbs-Merrill and the streamlined Knickerbocker Toy Company took over authorized doll production from the smaller-scale

Illustration, Raggedy Ann Stories *(1918), Johnny Gruelle.*

Illustration, Grimm's Fairy Tales *(1914), Johnny Gruelle.*

Georgene Novelties. The ripple effect of Gruelle's talent is very much with us today in the form of reprinted and derivative books and contemporary versions of his dolls and toys. Although the Gruelle family continues to be involved in all aspects of Raggedy publishing and licensing, history has proven that the 1960s marked the beginning of a highly diversified, much more nebulous merchandising era—a still-unfolding modern sequel worthy enough to be the subject of a future book.

Like the works of so many who contributed to the mainstream of American popular culture, Johnny Gruelle's creative and business legacy (and the primary sources that pertain) are spread among hundreds of libraries, museums, galleries, booksellers, family members, and private collectors, as perhaps they should be.

This far-flung nature of the resources made for a rousing and challenging treasure hunt. But more than once, when a tip seemed to lead nowhere or a fact could not be precisely confirmed or refuted, a rare doll or elusive document would somehow, as if by magic, fall squarely in my path. Such has been the alchemy and the synchronicity of this remarkable, magical journey.

I AM INDEBTED AND EVER GRATEFUL to the many friends, colleagues, and family members who assisted and encouraged me in all phases of my work. That group includes:

The Gruelle family, who shared information and materials and offered encouragement; namely, Jane Gruelle Comerford, Kit Gruelle, Kim Gruelle, Richard B. Gruelle, Ruth Gruelle, Terry Gruelle, Worth and Suzanne Gruelle, Megan Keating, Shane Keating, Will Murphy, Margaret Gruelle Owen, Peggy Y. Slone, and Tom and Joni Gruelle Wannamaker;

Friends and acquaintances of Johnny Gruelle, who shared materials and firsthand memories; namely, Lynda Oeder Britt, Jessie Soltau Corbin, Monica Borglum Davies, Helen Dayton, Betty MacKeever Dow, Mary R. Gill, Anne laDue Hartman Kerr, Janet Smith Leach, and Norman Meek;

The collectors, artists, librarians, curators, booksellers, scholars, and friends who provided information, materials, and valuable contacts; namely, Rollie Adams, Kim Avery, Beth Axelrod, Pauline Baker, Barbara Barth, David Begin, Burnette Benedict, Joel Cadbury, Charles Carton, John Cech, Sue Cloak, Ron and Sandy Crain, Nan Czyzewicz, Patricia Caze Deig, Jeanne Delts, Cynthia Dorfman, Barbara and Roy Dubay, John and Kathy Ellis, Flora Faraci, Betsy Fisher, Alice Fleming, Bruce Fox, Nan Galloway, Beverly Garcia-Garst, Sondra Gast, Kathy George, Elaine Goleski, Lynda Graves, Rebecca Greason, Peter Hanff, Rosemary and Durwood Hanline, Judy and James Hatch, Michael Patrick Hearn, Pen Jones, Joyce Link, Alison Hubbard and Bill Miller, Kathy Erskine Jenkins, Doris Knaus, Judd Lawson, Ellen Manyon, Jerry Marcus, Richard Marschall, Jeryl Metz, Craig McNab, Marge Meisinger, Peter Muldavin, John Murray, Carolyn O'Brien, Gayle O'Neal, Suzette Phillips, Pat Planton, Robert Plowman, Rick Richter, Beth Savino, Kayleen Seybrandt, Lawanda Smith, Patricia Smith, Mark Sottnick, Leonard Swann, Jr., Andrew Tabbat, Holly Tatson, Saundra Taylor, Andrea Telli, Gloria Timmell, Larry and Bonnie Vaughan, the Wardell family, Evon Webster, Helen White, Martin Williams, Barbara Young, Helen and Marc Younger, and Bob and Ruth Zimmerman;

The institutions and businesses that provided primary and secondary source material; namely, AP Wirephoto, Chicago Public Library, Children's Museum of Indianapolis, Cleveland State University Library, Fisher-Price, Inc., Holgate Toy Company, Indiana State Library, Indiana University Library, Indianapolis Public Library, Johnny Gruelle Raggedy Ann and Andy Museum, Joliet (Illinois) Area Museum, Library of Congress, Marshall Field's, Miami-Dade County Public Library, the Area Resource Center of the Nashville Public Library, National Archives, Franklin D. Roosevelt Library, Simon & Schuster, Sirocco Productions, Steiff Museum and Archives, Strong Museum, United States Copyright Office, University of Connecticut Library, University of Florida Library, University of Mississippi Library, and Vanderbilt University Library.

THE

GRUELLE IDEAL

It is the Gruelle ideal that books for children should contain nothing to cause fright, suggest fear, glorify mischief, excuse malice or condone cruelty. That is why they are called

"BOOKS GOOD FOR CHILDREN."

"The Gruelle Ideal," Johnny Gruelle Company, 1940s.

FINALLY, I AM EVER GRATEFUL to the following people whose extra help, vision, and generosity made *Raggedy Ann and More* the book it is; namely, Candy Brainard, Lauren Bufferd, Barbara Lauver, Brenda Milliren, Jacki Payne, Dan Scheffey, Caroline Cox Scheffey, and Dorothy Wortham, all of whom went to great lengths on many occasions to provide photographs, materials, and information; Susan Manakul, my expert research assistant, who located and accessed exactly what I needed precisely when I needed it; Joel Cadbury, Barry Cohen, Dolores Blyth Jones, Maggie Page, and James Summerville, all of whom offered advice on the manuscript; Rita Rosenkranz, my talented literary agent, who counseled with great warmth and insight; Dr. Milburn Calhoun, James Calhoun, Tracey Clements, Gwynn Harris, and Nina Kooij of Pelican Publishing Company, who understood my vision and transformed it into an elegant book; my extended California-and-Texas families, who faithfully and tirelessly cheered me on; and most of all, my husband Barry and sons David and Russell, whose patience, good humor, and unconditional love kept my spirits high and my heart open.

Illustration, John Martin's Book *(1918)*, *Johnny Gruelle*.

Raggedy Ann and More

Raggedy Ann and Raggedy Andy have cheered, comforted, and entertained multiple generations. (Beth Axelrod/Cynthia Dorfman and Michael Patrick Hearn/Worth and Suzanne Gruelle/Author's Collection)

A Dweller in Fairyland: Johnny Gruelle

I have always believed in encouraging the children in any beautiful belief that their imagination may inspire.

JOHN GRUELLE
1911

WHEN R. B. AND ALICE GRUELLE settled on John Barton as given names for their firstborn son, they likely had no idea that several decades before, the Englishman John Barton had distinguished himself as an outstanding maker of dolls. And they certainly had no way to know that their prophetically named baby, born on Christmas Eve 1880, would create his own very original, very American dolls.

Introduced as the country was emerging from World War I, the unassuming little cloth playthings that Gruelle named Raggedy Ann and Raggedy Andy heralded an era when a cleverly conceived character, if properly positioned and promoted, could launch and sustain an entire career for its creator.

Introduced in 1918 and 1920, respectively, Raggedy Ann and Andy not only set Gruelle on an irreversible career path; they became a merchandising phenomenon in the process. As dolls and toys, and as characters in newspapers, magazines, books, advertisements, songs, and films, Gruelle's rag dolls would be embraced by successive generations as symbols of kindness, cheer, and the very essence of childhood. However, the Raggedys were but two brush strokes on the expansive canvas of a man who, from the time he was a schoolboy in Indianapolis, wanted only to draw funny pictures.

Gruelle had his parents to thank, not only for artistic genes but also for the encouragement to put his own pen to paper. His father was an earthy, forthright artist who, though plagued by doubts over being self-taught, was outwardly proud, never

That struggle is inevitable should be a truth welcomed by the aspirant to success in art.

R. B. Gruelle
1904

apologizing for his roots. R. B. and Alice Gruelle headed a lively, bohemian household, rearing Johnny and his younger siblings, Justin and Prudence, to be tolerant, free-thinking, and most of all, passionate about pursuing the arts, whatever form that might take.

"There will always be a marked opposition among parents to having their boys choose art as a vocation," R. B. Gruelle wrote in the Indianapolis *Daily Sentinel* in 1904. "This is largely due not so much to art itself, but rather from the fear of possible failure, or a sense of the long and tedious struggle, the persistent effort, the experience of seemingly inevitable discouragement through and by which success is finally attained."

R. B. could well have been writing about his son John, a contemplative boy who spent his youth doodling on scratchpads and covertly coloring in the black-and-white illustrations in his fairy tale books, before casting his lot at age nineteen as a newspaper cartoonist. Working for a decade and a half in a profession in which the borrowing and sharing of ideas was the norm, young John Gruelle eventually focused his artistic eye and fertile imagination on fashioning a creation of his very own. When he did, his fanciful leanings and years in the trenches as a comic artist more than paid off.

What he chose to create was a doll—something to appeal to the hearts of children and the memories of adults. The stylized doll Gruelle named "Raggedy Ann" in his 1915 patent drawing evoked a sense of the past and a belief in the playful, the graphically superior design conveying a nostalgic whimsy and aesthetic appeal just right for its time. Appearing in every way to be hand-wrought, Raggedy Ann was a clever blend of the old and the new; the organic and the spiritual; the practical and the exuberant—a perfect doll for Americans then captivated by the sounds of ragtime and jazz and the aesthetics of the American Arts-and-Crafts Movement.

With a power and presence that were undeniable, the Raggedys became a national phenomenon almost overnight. Thanks to targeted marketing, word-of-mouth, and the dolls' own special charisma, by the mid-1920s the Raggedy Ann and Andy dolls and books were virtually everywhere. As

Clockwise: Johnny Gruelle at ages 4, 19, 35, and 49. (Margaret Gruelle Owen/ Norman Meek/Jane Gruelle Comerford)

(2)

Raggedy Ann and Raggedy Andy, P. F. Volland Company, early 1920s.

Johnny Gruelle as he saw himself, with revisions. (The Wardell Family)

the P. F. Volland Company continued selling the books, dolls, and related merchandise, the Raggedys brought their creator the recognition he both craved and deserved. Gruelle eventually became so well known that letters addressed simply "Johnny Gruelle, Raggedy Ann's House, Miami Beach" would reach him.

But Gruelle would come to chafe at being recognized *only* for his rag-doll characters. Throughout his career, he worked continuously as a magazine illustrator and comic artist, for both children and adults. And he had put forward a number of different doll and toy characters, as well as ideas for new books. But to his disappointment, Gruelle was rarely acknowledged for this other work, and none of his other characters would strike the fancy of a publisher, or the public, as Raggedy Ann and Andy had.

Johnny Gruelle's middle age (as it turned out, his final years) was spent quietly, working at his drawing board, fishing the south Florida waterways, and

Johnny Gruelle filled his stories, books, and correspondence with nonsensical critters such as this. (Richard Marschall)

When one is only two years old, he has very few sorrows.
Johnny Gruelle
1920

The publication of a new Raggedy Ann book was celebrated by the children who stood in line to meet the father of the Raggedys and receive their own special inscription on a book or doll. (Marshall Field's)

hobnobbing with friends and family over afternoon cocktails. At the time of his death, at age fifty-seven, Gruelle appeared to have made his mark, having obtained, thanks to the Raggedys, a sizable piece of the American Dream. But as friends and photographs confirm, he had bartered his health and his heart in the process.

The warmth in the clasp of a friendly hand is made of the magic of Sun-dust; whether it be the hand adorned with precious stones; the hand calloused with the friction of hardy toil; the foot of a dog; or the rag hand of a doll.

Johnny Gruelle
1920

Throughout his career, Gruelle had been no stranger to dissension and adversity. While blessed with a devoted wife, Myrtle, and two healthy sons (Worth, born in 1912, and Richard, born in 1917), Gruelle had lost his thirteen-year-old daughter, Marcella, in 1915. During the 1920s, his already strained relations with the Volland Company became fraught with misunderstandings and suspicion, and Gruelle began searching for a single new creation or other crowning achievement that could earn him a moniker besides "The Raggedy Ann Man."

By the 1930s, control of his intellectual property was a major issue for Gruelle, affecting his business dealings and outlook and ultimately involving him

Inspired by a photograph, Gruelle created this watercolor of his late daughter, Marcella. (Worth and Suzanne Gruelle)

(3)

P. F. Volland Company promotional booklet, 1920s. (Cox-Scheffey Collection)

in a protracted trademark infringement lawsuit. Yet, when given the chance, he had been ambivalent about launching the family-owned business that could have promoted and protected his creations and forestalled his legal problems.

Despite his yearnings and disappointments, Gruelle worked steadily and resolutely turning out his many works, even during the Great Depression. At the time of his death in 1938, sales of his Raggedy Ann and Andy books stood at well over 2,000,000, and more than 150,000 of his dolls had been sold along with assorted merchandise. During the decades that followed, Gruelle's creations would become an entire industry.

It is these many and varied works—some clearly derivative, others wildly original—that offer sharply focused reflections of who Gruelle *really* was: what he cherished and what he loathed; what dumbfounded him and what made perfect sense; what he fantasized about, what bored him, and what he thought would sell.

Had Johnny Gruelle enjoyed a longer life, he might never have created another set of characters to rival his ever-popular Raggedy Ann and Andy. And he may not have resolved his conflicting feelings about creativity, merchandising, and control. But, without a doubt, new and whimsical creations would have continued springing from the imagination of this gentle, playful man who knew that he had something quite special to give to the American public.

For a detailed account of the life and times of Johnny Gruelle, the reader is directed to the illustrated biography *Johnny Gruelle, Creator of Raggedy Ann and Andy*, by Patricia Hall (Pelican Publishing Company, 1993)

Treasures and Toys for Girls and Boys: Johnny Gruelle's Character Merchandise

Be it known that I, John B. Gruelle, a citizen of the United States, residing at New York City in the county of New York, and State of New York, have invented a new, original, and ornamental design for a doll.

JOHN B. GRUELLE
Design patent application
May 28, 1915

ON A NOVEMBER MORNING IN 1910, a young artist hurried through the doors of the imposing *New York Herald* headquarters, a sheaf of comic drawings tucked safely under his arm. The building's cornerstone owls blinked their red-lantern eyes inscrutably, witnessing, as they had many times before, the rather unremarkable beginnings of another young cartoonist's career.

Though a newcomer to New York, thirty-year-old John Gruelle was not without portfolio. With a decade's worth of daily and Sunday cartooning under his belt (done for a half-dozen Midwestern newspapers including the *Indianapolis Star* and *Cleveland Press*), he possessed more experience and confidence than many who had entered the portals of the well-regarded New York newspaper.

Gruelle had been summoned by the *Herald* from his home in nearby Silvermine, Connecticut, after winning a comic drawing contest that was, in reality, a recruitment scheme for a new Sunday cartoonist. Gruelle's entries, "Mr. Twee Deedle" and "Jack the Giant Killer," had been selected as first- and second-place winners from a pool of 1,500 entries. Extolling Gruelle's work for its artistry and appeal (but claiming they had never heard of him), *Herald* editors awarded him a cash prize—the tidy sum of $2,000—and a permanent position with the paper.

Gruelle's "Mr. Twee Deedle" was a full-page comic starring a professorial little woodsprite inspired by a Silvermine neighbor's storytelling. In addition to syndicating Gruelle's prize-winning strip, *Herald* editors positioned it on the front page of its Sunday comic section (replacing Winsor McCay's popular "Little Nemo" in the process). Here, Gruelle's sage little elf could present his clever combination of traditional and contemporary virtues that were guaranteed to delight parents and sell Sunday newspapers.

Years before Raggedy Ann made her official debut, Gruelle was dressing subjects of his satire in rag-bag attire: striped leggings, flower-decked hats, and oversized shoes. The <Indianapolis> Sentinel, *March 31 and June 19, 1904; the* Cleveland Press, *May 28, 1907.*

(4)

SECTION 5.

SECTION 5.

The Courier-Journal.

Louisville, Ky. Sunday.

JUNE 18 1905

BERNIER — DINK — ED. A. GOEWEY — C.H. WELLINGTON — JOHN GRUELLE.

OUR FUNNY SECTION'S HAPPY FAMILY WILL HAVE AN EXCITING FOURTH.

Gruelle was one of several World Color Printing artists who collaborated on this 1905 Independence Day greeting. (David Begin)

Sensing the commercial potential of a character that embodied both old-fashioned and street-wise traits, the *Herald* had moved full-speed ahead with plans for a doll based on the Mr. Twee Deedle character. While some (including, perhaps, Gruelle) may have had their doubts about trying to capitalize on a character doll based on a yet-to-be published comic page, the *Herald* was convinced of Mr. Twee Deedle's appeal and stood ready to enter the playthings fray.

By 1911, the licensing and selling of popular characters had become big business, though by no means risk-free. Specialty catalogues devoted full-page spreads to dolls, toys, and merchandise depicting comic and book characters. Touted in trade magazines such as *Playthings* and *Toys and Novelties* and showcased at the annual New York City Toy Fair, character playthings enjoyed a caché and marketing edge that were unparalleled, each incarnation serving to promote related items.

Florence Upton's Golliwogs, Palmer Cox's Brownies, and the multi-sized Fairyland Rag Dolls—all based on characters in illustrated children's books—had paved the way. Furthermore, dolls linked to the newspaper "funny pages" were proving even more popular. Consumers seemed especially willing to spend hard-earned cash on recognizable dolls based on characters they could "visit with" every Sunday morning in the newspaper.

continued on page 31

Though most of his early cartoons were for adults, Gruelle by 1909 was creating features for and about children, such as this one done for the Cleveland Press, *November 14, 1909.*

ARTISTS!

Here Is a

Prize of $2,000!

The NEW YORK HERALD, to obtain pictorial features as dainty as Fluffy Ruffles and as dashing as the Widow Wise, offers

A Prize of $2,000

for the best series of five full page cartoons submitted in open competition.

All artists, amateur or professional, may enter this contest, and the choice of subjects is unrestricted. The subject chosen by the artist, however, must be adhered to throughout the series submitted.

The prize winning series is to be the HERALD'S property, and it reserves the right to continue the series if conditions warrant.

The HERALD does not bind itself to purchase any of the cartoons submitted, but, by mutual agreement, may buy any series entered in the competition.

The contest is open NOW and will close on NOVEMBER 1, 1910.

Competitors should address the Cartoon Contest Editor, New York Herald office.

This small announcement, which appeared in fall of 1910 in the New York Herald, *forever changed the course of Johnny Gruelle's career.*

(5)

MAYBE You Don't Know the Comical Little Chap, but the Children and Every One Else Are Bound to Like Him When He Appears in the Funny Pictures

(Copyright, 1911, by the New York Herald Co. All rights reserved.)

WHEN children expect company with whom they have never played they wonder, as a rule, what the new playmate will be like. So it is only just to the wishes of the children that I introduce to them Mr. Twee Deedle, who will appear for the first time in the pages of this newspaper next week. It might be that some children may happen to know a Twee Deedle, for I am sure there are more Twee Deedles than the one with whom I happened to become acquainted.

The Twee Deedle I know lived in an old, gnarled apple tree growing in a thick woods out near Silver Mine, Connecticut.

Mr. Gutzon Borglum, the sculptor, who owns the property, discovered the tree and the path leading to the little hole which the Twee Deedle uses as a doorway. Mr. Borglum often told his little boy and girl about the Twee Deedle and showed them the path and doorway; but the children could never get a peep at the Twee Deedle, even though they watched the doorway for hours at a time.

When I moved to Connecticut and heard my little girl speak of the Twee Deedle I could not imagine what she meant, so did not give the matter much thought.

That afternoon the Borglum children took my little girl to the tree, and she came running all excited.

"Papa," she cried, "there really must be a Twee Deedle living in the old tree, for there is a little path that leads down to the spring, and there is the cutest little hole in the tree that I'm sure he uses for a doorway."

So I put on my hat and we went through the woods to the tree. Sure enough, there was the path, just about three inches wide, winding around little sticks and stones just like a grown-up man's path, and leading to the crystal spring bubbling about ten feet from the tree. We brushed the ferns and flowers aside, and there was the doorway, just large enough for a tiny little person to enter.

I have always believed in encouraging the children in any beautiful belief that their imagination may inspire, so I told my little girl that I guessed it must be a Twee Deedle that lived in the old tree, even though I thought the path was made by a little squirrel or a chipmunk.

However, I thought of the place quite often, and while wandering around through the woods I would often sneak up to the tree and watch for the chipmunk that I thought had made the path.

I brought acorns and placed them in the path, but when I would look the next time the path would be clear and the acorns would be in the grass at the side, so I knew whatever lived there was not a squirrel, or he would have eaten the nuts.

One day I left a piece of candy. It was gone the next time I looked. Apples that I placed in the path would be rolled off to one side when I would look the next day.

One evening I took a large apple and covered it with white paint and placed it just in front of the little doorway; then I went away. Early the next morning I found the apple in the grass at one side of the little path, and the paint was covered with prints of a little hand no larger than a tiny kitten's paw.

There was now no doubt in my mind that the Twee Deedle really lived in the tree, but how to see him was the question that bothered me.

I found out that he would take the candy away which I placed in the path, so I decided to catch him in a rat trap baited with candy. I went to town and bought a large trap and placed it in the path in the evening. I could scarcely sleep that night, and when the clock struck two I was wide awake, so I dressed and walked through the woods. There was no moon that night, and it was quite hard to find the tree. I do not think I should have found it at all had there not been fifty or one hundred lightning bugs on the ferns above the little path. As I carefully crawled up to the tree I heard the trap spring shut with a loud snap and the lightning bugs all flew up in the air and then settled again on the ferns. The light from the fireflies cast a soft yellow glow for three or four feet around the place, so I could see the Twee Deedle sitting inside the trap eating the candy.

I wish that all you children could have seen him. Just imagine a little figure only eight or nine inches high, with a queer, comical shaped hat looking like a morning glory, only the brim was scalloped more like the petals of a daisy. His head was just about the size of a dollar, rather large for his body, and his hair, which was parted in the middle, Indian fashion, fell down over his shoulders. He wore a pointed collar, and at each point there hung a round button. His shirt coat, or dress, scarcely reached to his knees, and was cut in points just like his collar. He wore tights that fitted his legs so snugly that it seemed as if they were a part of him, and his feet were encased in the cutest little moccasins that you could imagine.

He was the most cunning little figure I had ever seen, as he sat cross-legged, tailor fashion, in the bottom of the trap and ate the candy.

I think he must have heard my breathing, for he looked up in a jerky way, and in a little teeny weeny voice asked:—

"That you, Mr. Fox?"

I didn't know what to say. I didn't care to frighten him, but when he asked again, "That you, Mr. Fox?" I said:—

"No; this is Johnny Gruelle."

"Never heard of you before," he said, and went on eating the candy. He wasn't a bit afraid.

"I brought you the candy," I replied, by way of opening the conversation, "and I put the trap there to catch you in."

"What trap?" he asked, with his mouth full of candy.

"Why, the trap you are sitting in, of course," I replied.

He gave a little tinkly laugh.

"This isn't a trap," he said, and he stood up and walked right through the wires, just as if there were none there, and then walked back into the trap and sat down again.

"I thought that was a good trap," I said. "Mr. Hubble, the hardware man, said the wires were good and strong and would hold anything."

"Well, he has never tried to hold me," said the Twee Deedle.

Well, it would make quite a long story if I would tell you all we talked about, but I asked him everything I could think of and he didn't seem to mind my inquisitiveness.

I told him I was trying to find an idea for a comic series so that I could enter the prize competition, and he told me that if I would send the drawing he suggested I would win the prize. So he told me just what to make, and I made it.

Every morning last autumn at about two o'clock I went over to the tree and we had long talks, and he told me of different adventures he and Dickie had gone through. He didn't tell me who Dickie was or where he lived, so I cannot tell you any more of Dickie than I tell you in my pictures.

With the coming of cold weather I discontinued my early morning visits to the Twee Deedle tree, and last week when I went there the snow had banked up over the little doorway and covered the path. I scraped it away and called, but received no answer, so I do not know what has become of Mr. Twee Deedle. Perhaps he goes South each year, as the birds do, or perhaps he goes away up into the trunk of the tree and in a little feathery bed sleeps until Spring knocks at the door and bids him awaken.

I will ask him next summer, if he is in the old apple tree, just where he goes each winter, and perhaps I shall let you into the secret.

JOHN B. GRUELLE.

Gruelle's playful "Mr. Twee Deedle" was introduced to readers on January 29, 1911.

The foulard tie, short pants, and striped legs of Gruelle's 1908 "Bud Smith" comic character foreshadowed his Raggedy Andy character.

Johnny Gruelle's Character Dolls

This table summarizes the manufacturing history of character dolls created or directly inspired by Johnny Gruelle. Measurements (to the closest inch) are to top of head; however, actual sizes can vary, depending on fabrication. Dates refer to manufacturing period, which did not necessarily coincide with license duration or retail availability. Doll identifiers that appear in quotation marks are either commonly used terms or the author's nomenclature, not official company designations. They are presented here to assist in identifying and categorizing the dolls.

Company/ Purveyor	Doll	Size(s)	Distinguishing Features	Production Dates	Manufacturer
Steinhardt & Bro.	Twee Deedle	18", large (size unspecified)	felt & composition; cloth trademark tag	1911-	Steinhardt
Margarete Steiff	Mr. Twee Deedle	14", 20"	all felt; center-seamed face (for 20" doll, only prototype documented)	1913-15	Steiff
Johnny Gruelle	Raggedy Ann ("Trademark")	unknown	trademark tag; no dolls documented	1915-	Johnny Gruelle
P. F. Volland Company	Raggedy Ann ("Cottage")	15-18"	hand-painted face & shoes; dress fabric matches Raggedy Ann Stories cover (later dolls had cloth feet & varied dress fabric)	1918-19	Non-Breakable
P. F. Volland Company	Raggedy Ann ("Patent")	15-16"	patent stamp; machine-stamped "wistful" face	1920-26	Muskegon
P. F. Volland Company	Raggedy Andy ("Crescent Smile")	16"	crescent smile; large-scale head & hands	1920-27	B-K-B; Beers
P. F. Volland Company	Raggedy Ann & Raggedy Andy ("Oversized")	30-36"	Anns (1924-25, 1928) have line smiles and outlined noses; Andys (1920-21) have hand-painted faces with line smiles & unoutlined noses; Andys (1924-25, 1928) have crescent smiles and unoutlined noses	1920-21; 1924-25; 1928	B-K-B; Beers
P. F. Volland Company	Little Brown Bear ("Ross & Ross")	12"	plush head & paws; felt pants; cloth trademark tag	1921-23	Ross & Ross
Woman's World	Johnny Mouse	10"	felt & cotton; sewn-on clothing; string tail	1921-22	Ross & Ross
P. F. Volland Company	Eddie Elephant ("Ross & Ross")	14"	felt & cotton; made to stand; hand-painted face; flat-bottomed feet; cloth trademark tag	1922-23	Ross & Ross
P. F. Volland Company	Sunny Bunny ("Ross & Ross")	12"	v-shaped stitched nose; brown plush head and paws; red felt pants; green felt vest/coat; no top hat documented; cloth trademark tag	1922-23	Ross & Ross
P. F. Volland Company	Beloved Belindy	15"	hand-painted or silk-screened face; solid red shoes, & bandana	1926-33	Unknown
P. F. Volland Company	Raggedy Ann & Raggedy Andy ("Single Eyelash")	14-16"	resembles Gruelle's later book illustrations; lower lash a line, dot, comma, or inverted triangle; line smile; Ann & Andy not a matched set	mid-late 1920s	Moore &/or Gerlach-Barklow
P. F. Volland Company	Raggedy Ann & Raggedy Andy ("Transitional")	15-16"	Ann has heavily blushed cheeks, low-set line smile, unoutlined nose; Andy has heavy eyebrows, line smile, unoutlined nose; Ann & Andy not a matched set	late 1920s-early 1930s	Moore &/or Gerlach-Barklow
P. F. Volland Company	Beloved Belindy ("Oversized")	30-36"	unknown (advertised, but not documented)	1928	Unknown
P. F. Volland Company	Raggedy Ann & Raggedy Andy ("Finale")	14-15"	bold black eyebrows; paper hang tags; Ann has outlined nose, Andy has unoutlined nose; later Andys have gingham instead of plaid shirts; Ann & Andy are a matched set	1931-33	Moore
P. F. Volland Company	Little Brown Bear	12"	felt head, paws & feet; fabric pants; no cloth body tag	1931-33	Unknown
P. F. Volland Company	Eddie Elephant	11"	felt & cotton; made to sit; sewn/embroidered features; center-seamed feet; no cloth body tag	1931-33	Unknown

Company/ Purveyor	Doll	Size(s)	Distinguishing Features	Production Dates	Manufacturer
P. F. Volland Company	Sunny Bunny	12"	horizontal stitched nose; auburn plush head & paws; green cotton pants; red felt coat; yellow felt vest; black top hat; no cloth tag	1931-33	Unknown
P. F. Volland Company	Percy <the> Policeman	15-16"	center-seamed face; same body as Uncle Clem; felt-backed shoe-button eyes; no cloth tag	1931-33	Unknown
P. F. Volland Company	Uncle Clem	15-16"	center-seamed face; same body as Percy; felt-backed shoe-button eyes; no cloth tag	1931-33	Unknown
P. F. Volland Company	Pirate Chieftan	17"	center-seamed face; long nose; yarn beard; small shoe-button eyes; no cloth tag	1931-33	Unknown
P. F. Volland Company	Eddie Elf	15"	flat felt head; small shoe-button eyes; pink felt hat; no cloth tag	1931-33	Unknown
Exposition Doll & Toy Manufacturing Company	Raggedy Ann	18"	silk-screened face; burgundy yarn hair; removable felt shoes; cloth trademark tag on dress hem; made to sit	1935-36	Exposition
Molly-'Es Doll Outfitters*	Raggedy Ann & Raggedy Andy ("Archtype")	18", 22"	Ann & Andy have identical body & face designs; silk-screened faces; printed heart & logo on chests (most dolls); multi-color striped legs; blue feet; 22" dolls have smaller-scale heads	1935-37	Molly-'Es
Molly-'Es Doll Outfitters*	Raggedy Ann & Raggedy Andy ("Mollye Babies")	14"	identical body & face designs; yarn hair "fringe" attached at hairlines; fabric head backs; some made to sit	ca.1935-37	Molly-'Es
Molly-'Es Doll Outfitters*	Beloved Belindy	unknown	(only prototype documented)	ca. 1935	N/A
American Toy & Novelty Manufacturing Company*	Buddy & Sis ("Huggable Nursery Pets")	15-16"	oil-cloth faces; design of matched sets changed over time; multi-fabric bodies and clothes; cloth tags	ca. late 1930s-1950s	American
Georgene Novelties	Raggedy Ann & Raggedy Andy ("Outlined Nose")	19", 23", 31"	outlined noses; identical body & face designs; wool yarn hair; red-and-white striped legs; printed cloth body tags (red printing for Ann, blue printing for Andy)	1938-early 1940s	Georgene
Georgene Novelties	Beloved Belindy ("Standard")	18-19"	outlined nose; medium-to-dark brown cloth body; solid red (with white polkadots during WWII) shoes, bandana, blouse; four buttons on blouse; shirt-button eyes until 1946, shoe-button eyes thereafter; identifying stamp on head or bottom	ca. 1939-late 1940s	Georgene
Georgene Novelties	Raggedy Ann & Raggedy Andy ("Awake-Asleep")	12-13"	awake faces on one side, asleep faces on other; identical body & face designs; outlined noses until 1944; unoutlined noses thereafter; printed cloth body tags	ca. 1940-46	Georgene
Georgene Novelties	Camel with the Wrinkled Knees	10" high	light-brown felted flannel; made to stand; celluloid "googly eyes"; yellow/blue/red blanket; white tin buttons; no company markings	1941-early 1950s	Georgene
Georgene Novelties	Raggedy Ann & Raggedy Andy ("Wartime")	19", 23", 31"	unoutlined noses; identical body & face designs, cotton-yarn hair; feet & legs of varied print fabrics; printed cloth body tags	ca.1943-46	Georgene
Georgene Novelties	Beloved Belindy ("Little")	15"	pale-brown cloth body; outlined nose; muted clothing colors; shirt-button eyes; four buttons on blouse; no company markings or body tags	mid-1940s-late 1940s	Georgene
Georgene Novelties	Raggedy Ann & Raggedy Andy ("Postwar/Silsby")	15", 19", 23", 31"	"Silsby" on printed cloth body tags; unoutlined noses; identical body & face designs; A-line dress (Ann); white shirt collar (Andy)	1946-ca. 1950	Georgene
Georgene Novelties	Beloved Belindy ("Forward Feet")	15"	front-facing boxy feet; medium-brown cloth body; unoutlined nose; red blouse, shoes, bandana; shirt-button eyes; two snaps on blouse; no company markings	late 1940s-1950s	Georgene
Georgene Novelties	Raggedy Ann & Raggedy Andy ("Fifties")	15", 19", 23", 31", 45"	curvaceous facial designs; identical body & face designs; unoutlined noses; metal-disc eyes (plastic on 45"); blue ribbon apron strings (Ann); printed cloth body tags	ca.1950-early 1960s	Georgene
Georgene Novelties	Raggedy Ann & Raggedy Andy ("Plain Tag")	19", 23", 31"	rubber-stamped white or flesh-colored body tags; identical body & face designs; curvaceous faces	ca.1962	Georgene

2—So off into the air they go, a-floating up and down,
And out across the lofty spires and buildings of the town.

4—The dolls all come to life at once, and eagerly they ran
To Dolly, whom they seize and hug as quickly as they can.

6—And down a helter-skelter there a merry ride they take,
But, striking at the bottom, all the little doll heads break.

Gruelle used the theme of "come-alive" dolls in his December 22, 1912, episode of "Mr. Twee Deedle."

Myrtle and Marcella Gruelle, Silvermine, Connecticut, 1911. (Tom and Joni Gruelle Wannamaker)

continued from page 25

On February 3, 1911 (five days after the first "Mr. Twee Deedle" page appeared), the *New York Herald* authorized A. Steinhardt & Bro. to begin manufacturing a trademarked character doll called "Twee Deedle," which was advertised in *Playthings* just in time for the 1911 New York City Toy Fair. By 1913, the *Herald* had reassigned doll rights to Margarete Steiff, a German company famous for its felt character dolls, and a Steiff Mr. Twee Deedle doll was made available in the U.S. and abroad.

Judging from the scarcity of extant examples, production and sales for the <Mr.> Twee Deedle dolls were probably less than brisk. Despite this (and the fact that he was given little remuneration and virtually no credit for the dolls' design), Gruelle had learned a precious lesson, one that would come to drive and define his entire career—namely, the power of character merchandising.

Gruelle continued chronicling Mr. Twee Deedle and his life-size proteges, Dickie and Dolly, alternating the storylines between make-believe fairy-tale adventures and up-to-date social issues such as homelessness and urban blight. When Gruelle was approached in 1915 about creating an animated film

The news that the adventures of "Mr. Twee Deedle" will be dwelt upon each week in this New York paper, as well as in newspapers all over the country, will certainly have a strong effect upon the popularity of this doll.

Playthings
March 1911

JOLLY LITTLE JOKERS: THE TWEE DEEDLE DOLLS

The first commercial dolls based on Gruelle's Mr. Twee Deedle character were manufactured by A. Steinhardt & Bro. beginning in February 1911. Available in a small, 18" version as well as a larger unspecified size, the dolls were set to retail for $1.00 and $1.50. Made with felt bodies and Steinhardt "Neverbreak" composition heads, the Steinhardt Mr. Twee Deedle came dressed in green pants, yellow shoes with red pom poms, a yellow coat (with two large red pom poms, white pointed collar with bells on each point, and white turned-back cuffs), and an orange harlequin hat. On the hem of each doll's coat was sewn a cloth label with trademark information and the name "Mr. Twee Deedle."

In 1912, the 18" Steinhardt doll was featured in Gimbel's catalogue; however, its facial design and clothing differed from the Steinhardt dolls advertised the year before. Gimbel's mistaken claim that the Mr. Twee Deedle dolls possessed "Can't Break 'Em" heads (a trademarked feature of Horsman dolls) rather than Steinhardt's "Neverbreak" heads has led to the erroneous conclusion that Horsman produced and/or assembled Mr. Twee Deedle dolls.

The second commercial doll based on Mr. Twee Deedle was produced by Margarete Steiff. Referred to as "Amerikanische Karikatur" (American Comic) in a 1913 company catalogue, the Steiff Twee Deedle doll measured 14" and was priced to retail in the U.S. at $1.70. With the center-seamed face for which Steiff dolls are famous, this Twee Deedle had hair and small button eyes and came dressed in yellow tunics with puffed red sleeves, a white Vandyke collar, and a yellow-and-red pointed hat.

Particular description of goods.—Refrigerators.
Claims use since Feb. 1, 1911.

Ser. No. 54,723. (CLASS 22. GAMES, TOYS, AND SPORTING GOODS.) A. STEINHARDT & BRO., New York, N. Y. Filed Feb. 24, 1911.

TWEE DEEDLE.

Particular description of goods.—Dolls.
Claims use since Feb. 3, 1911.

A trademark was granted to Steinhardt & Bro. seven months after its Twee Deedle doll was introduced to toy buyers. Advertising sternly warned the trade to beware of unlawful imitations made in violation of its trademark and the New York Herald's *copyright.*

(6)

(7)

Playthings *predicted the new Twee Deedle doll had "every chance of becoming as well known and liked as Little Nemo, Foxy Grandpa, Buster Brown or any other of their illustrious predecessors." (Top) Gruelle's rendering (the* New York Herald, *April 30, 1911), (bottom; left to right) 1911 Steinhardt Twee Deedle (*Playthings, *March 1911) and 1912 Steinhardt Twee Deedle (Gimbel's catalogue, 1912).*

The 14" Steiff Twee Deedle (which, according to company sources, sold only modestly in the U.S. and Britain) was followed two years later by a 20" version. However, because of World War I and the resulting trade restrictions with Germany, this larger doll may never have gotten beyond the prototype stage.

(8)

Prototypes of the 14" (top) and 20" (bottom) Steiff Mr. Twee Deedle dolls. (Steiff Museum, Margarete Steiff GmbH)

Gruelle's Christmas greeting, John Martin's Book, *December 1914.*

version of the syndicated "Mr. Twee Deedle" comic page, he received the blessings of *Herald* publisher James Gordon Bennett to go forward with the project. Regrettably, a "Mr. Twee Deedle" cartoon never materialized, but if it had, it would have been among the first commercial animated cartoons ever released.

Meanwhile, Gruelle was taking on other illustrating commissions: for the Arts-and-Crafts-inspired children's magazine *John Martin's Book,* and the juvenile book publisher Cupples & Leon, among others. As he immersed himself in his comics and illustrations, Gruelle became more and more attuned to the world of children. He sought out and used both old-fashioned and contemporary themes he knew youngsters would like, creating images and episodes overflowing with the lessons, the adventures, and the playthings of childhood.

In his earliest "Mr. Twee Deedle" episodes, Gruelle occasionally incorporated a floppy-limbed rag doll with scraggly hair. Though often nameless, this tattered little doll (usually shown trailing from the hand of her human mistress Dolly) sometimes went by the name "Rags" or "Lou." Despite her worn, faded appearance, she was a bright little beacon of what lay ahead for John Gruelle.

Just what that was became clear on May 28, 1915, the day Gruelle submitted an original design to the U.S. Patent Office. His detailed pen-and-ink drawing showed front and side views of a cloth doll dressed in a plain frock, with puffy peplum-style overskirt and white eyelet pantalettes, and beflowered straw hat tied under her chin. With her stringy hair, outstretched arms, wide grin, and round, three-dimensional eyes, she was the merry antithesis of a prim wasp-waisted European-style doll. The design for "Raggedy Ann" (whose name was neatly printed

THE CUPPLES & LEON BOOKS

In 1913, the New York-based Cupples & Leon company published *Mr. Twee Deedle,* a reprint booklet of Gruelle's *New York Herald* comic pages. In 1914, the company issued a second reprint, *Mr. Twee Deedle's Further Adventures,* and that same year offered Gruelle his first major book-illustrating contract to produce a set of color plates and pen-and-ink drawings for a hardcover edition of *Grimm's Fairy Tales.*

This ambitious project was followed by commissions to illustrate Hector Malot's *Nobody's Boy* and six books in Cupples & Leon's "All About" storybook series, published in 1916 and 1917. In all, Cupples & Leon would publish fourteen titles illustrated and/or written by Gruelle, who never forgot the Christmas when he was invited to the company's showroom to pick out dozens of free books for himself and as gifts.

(9)

(10)

Grimm's Fairy Tales *(1914) and the abridged* Grimm's Fairy Stories *(1922) that featured only three of Gruelle's original dozen color plates.*

(13)

(11)

(12)

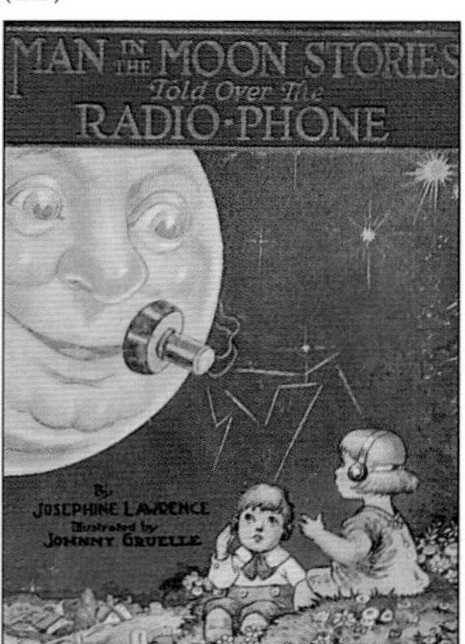

Gruelle illustrated Hector Malot's Nobody's Boy *(1916) and Josephine Lawrence's* Man in the Moon Stories Told Over the Radio-Phone *(1922) for Cupples & Leon.*

(14)

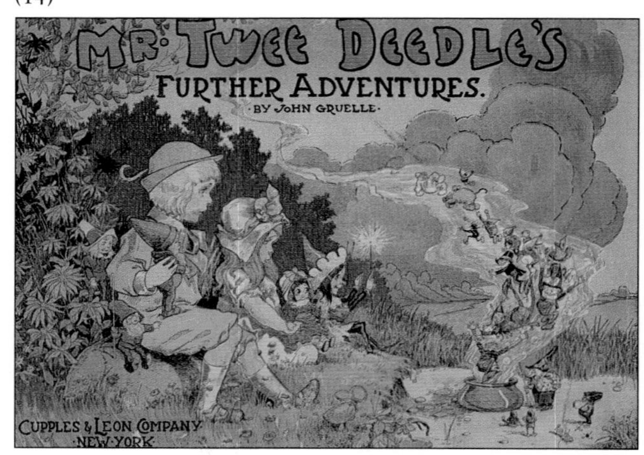

Cupples & Leon's Mr. Twee Deedle *(1913) and* The Further Adventures of Mr. Twee Deedle *(1914). Only the second volume featured Gruelle's name on the cover. (Homer Babbidge Library, University of Connecticut/Jacki Payne)*

In 1956, Cupples & Leon sold its interests to Platt & Munk, which for several years continued issuing *The All About Story Book*, a 1929 Cupples & Leon compendium that included five of Gruelle's "All About" titles.

(15)

(16)

(17)

(18)

(19)

Gruelle's "All About" books (1916 and 1917) were sold singly and in boxed gift sets. The All About Story Book *(1929) included five Gruelle titles.*

on her hat ribbon) was a synthesis of the many folksy little dolls that Gruelle had, for years, been including in his comics and illustrations.

The design patent for Raggedy Ann was granted on September 7 for a period of fourteen years. Meanwhile, John Gruelle had forged ahead, fashioning and fabricating multiple Raggedy Ann dolls. The exact number is not known, but there were enough dolls made to sell through interstate commerce. That qualified for a "Raggedy Ann" trademark, which was registered on November 23, 1915. With both a design patent and trademark to her name, Gruelle's little rag doll had staked her claim as a well-protected, irrefutably commercial plaything.

The time was right for a doll such as Raggedy Ann. Many American and English companies already had turned to cloth dolls, after observing that both boys and girls preferred sturdy play dolls to fragile bisque or composition creations. By 1915, the

A doll of any hard substance is such a long, long way from the ideal. A more or less yielding body is an essential to the perfect doll all designers are striving toward.

Kate Jordan
Toys and Novelties
April 1914

E. I. Horsman Company, Borgfeldt, Dean's Rag Book, and Martha Chase, to name only a few, were offering complete lines of fabric dolls with hand-painted and lithographed faces that sold in stores such as Gimbel Brothers, Macy's, and F.A.O. Schwartz.

Crediting the popular Golliwogs for igniting public interest in cloth character dolls, the trade magazine *Toys and Novelties* reported: "Rag baby dolls and character dolls are a prominent feature of the doll trade and the latter especially has provided some sensational features for the coming season."

It stands to reason that Gruelle might have used some of his homemade dolls as samples to stimulate the interest of a commercial doll manufacturer. Companies such as Borgfeldt (a thriving importer, assembler, and distributor that had handled both the Steinhardt and Steiff Twee Deedle dolls), as well as doll manufacturers such as Horsman or Krueger (all of which were actively searching for new marketable

continued on page 43

Peerless Equipment Company of New York enlisted Gruelle to illustrate this sheet music satirizing the escalating political situation in Europe. (Kim Gruelle)

Gruelle's color plates for Grimm's Fairy Tales *were formal portrayals of the lissome maidens, noble knights, and skulking common folk that peopled the Grimms' tales.*

The wee folk in Gruelle's March 15, 1913, cartoon for Judge *magazine were forerunners to the gnomes and "Dwarfies" that would appear in his later work.*

UNITED STATES PATENT OFFICE.

JOHN B. GRUELLE, OF NEW YORK, N. Y.

TRADE-MARK FOR DOLLS.

107,328. Registered Nov. 23, 1915.

Application filed June 17, 1915. Serial No. 87,361.

STATEMENT.

To all whom it may concern:

Be it known that I, JOHN B. GRUELLE, a citizen of the United States of America, residing in New York city, New York, and doing business at No. 210 West One Hundred and Eighth street, in the borough of Manhattan of said city, have adopted for my use the trade-mark shown in the accompanying drawing, for dolls, in Class 22, Games, toys, and sporting goods.

The trade mark has been continuously used in my business since June 5th, 1915.

The trade mark is applied to the dolls by affixing a label bearing the trade mark.

JOHN B. GRUELLE.

RAGGEDY ANN

DECLARATION.

State of Connecticut county of Fairfield ss:

JOHN B. GRUELLE, being duly sworn, deposes and says; that he is the applicant named in the foregoing statement; that he believes the foregoing statement is true; that he believes himself to be the owner of the trade mark sought to be registered; that no other person, firm, corporation or association, to the best of his knowledge and belief, has the right to use said trade mark, either in the identical form or in any such near resemblance thereto as might be calculated to deceive; and that the said trade mark is used by him in commerce among the several States of the United States; that the drawing presented truly represents the trade mark sought to be registered; and that the specimens shown the trade mark as actually used upon the goods.

JOHN B. GRUELLE.

Subscribed an dsworn to before me this 9th day of June, 1915.

[L. S.] CHESTER S. SELLECK,
Notary Public.

Copies of this trade-mark may be obtained for five cents each, by addressing the "Commissioner of Patents, Washington, D. C."

In contrast to the more generic cloth dolls whose designs filled the U.S. Patent Gazette, Gruelle's Raggedy Ann was a skillfully rendered, quite distinctive character. Gruelle claimed that the logo "RAGGEDY ANN" "has been continuously used in my business since June 5, 1915" and that it "is applied to the dolls by affixing a label bearing the trade mark."

Gruelle's boxed Mrs. Cluckluck Stories *unfortunately never got beyond the prototypes shown here. (Worth and Suzanne Gruelle)*

A DOLL NAMED RAGGEDY ANN: INSPIRATIONS AND INFLUENCES

One of the most imitated dolls in history, Raggedy Ann has inspired her share of doll makers and designers who have freely borrowed elements of her appearance, persona, and even name in hopes of coming up with a doll of their own—something to rival the whimsy, the grace, and the commercial potential of Johnny Gruelle's little rag doll. Raggedy Ann, likewise, was not created in a vacuum. There were inspirations aplenty to feed Gruelle's imagination as he conceptualized a soft, amusing rag doll that he hoped would sell.

By all accounts there had been a homemade cloth doll in Gruelle's family, owned by his mother, Alice, when she was a child growing up in Illinois. Supposedly, this doll was passed along to Johnny's sister Prudence, some time before 1900. Some time after 1900, Johnny discovered it in his mother's Indianapolis attic. After treating the faded doll to a comical new face, he gave it to his young daughter, Marcella. Over time, Gruelle became convinced that a new doll like the old family one that so captivated his little girl could be a marketable item.

As a cartoonist and freelancer, Gruelle was certainly privy to good ideas and inspirations during the

Gruelle's fascination with the white faces and striped legs of circus clowns and stage characters such as Fred Stone (here as the Scarecrow in the 1902 musical "The Wizard of Oz") was reflected in his design for the loppy, floppy Raggedy Ann.

THE RAG DOLLY

The little china clock in the nursery had just struck 12, and Bessie was sound asleep in her bed. There was the queerest noise in the doll house in the corner of the nursery, and Bes-

sie's kitten that had been sleeping beside her got up to see what was the matter.

The beautiful new bisque doll Arabella was telling the old rag dolly Rags how a little boy had poked one of her eyes out. Rags told her how sorry she was. But Arabella just shook her golden curls and said, "You shouldn't feel sorry for me, for just see what beautiful clothes I have and how pretty I am, and you have only old clothes and no yellow curls."

This hurt Rags' feelings very much, for even if she didn't have pretty clothes she was sweet and kind, but she only said, "Yes, you are perfectly beautiful."

But the kitten never had liked Arabella since she had fallen on him one day, so he said, "Rags can play outdoors and help Bessie make mud pies, and don't have to worry about spoiling her dress. Often she has fallen out of the doll buggy and it hasn't hurt her a bit."

Arabella knew she always had to take care of her pretty clothes, and when she had fallen out of the buggy she had broken one of her arms. But she said, "Anyway, I know Bessie likes me better because she takes me with her to the park."

This made Rags cry, and the kitten rubbed against her, purring to her not to cry.

The next morning when Bessie was ready to go to the park, she picked up Rags, saying to her mamma, "I guess I'll take Rags today; she is nicer to play with because her dress is old and she don't break when she falls."

As happy as Rags was over this, she felt sorry for proud Arabella because she had to stay at home.

Raggedy Dolls for Little Mothers

By Laura B. Starr

GIVEN the raw material and a little preliminary instruction from mother or teacher, and it will soon be found that doll-making has become more than a source of temporary amusement to the child. The desire to love and be loved is manifest in most girls from the moment the first doll is clasped in their arms. The mothering instinct is very strong—if by chance this is not the case, it should be cultivated.

The finding of many grotesque rag dolls in Egyptian tombs, prove that rag dolls have been first favorite from time immemorial. These funny little babies were laid in the dead child's arms by the ancient Egyptian mother in the expectation that they would cheer and comfort the little ones in the next world as they had done in this.

MADE BY A GIRL TEN YEARS OLD

The faces are painted, each true to its type, and the clothing is so arranged that by a gentle flop, a dexterous turn of the wrist, the skirts fall into place and you one moment have a white doll, and the next a black Dinah. The surprise of the small child in seeing the transformation is complete and most amusing. Indeed, I have seen older children much interested in the funny little mannikin.

The American rag doll is typical of the home-made dolls of fifty years ago. Her face and features are marked with oil paint, but having been the beloved plaything of two generations, she looks a bit the worse for wear, but nevertheless she is still good for another generation. There is a woman over eighty years of age near Boston who has turned her clever fingers to profitable account by making

(Top) The Cleveland Press, *June 7, 1909; (bottom)* McCall's Magazine, *February 1912.*

time he was envisioning his Raggedy Ann. In June 1909, a fellow cartoonist at the *Cleveland Press* (who signed his name "Valentine") published an illustrated children's story called "The Rag Dolly," starring a cloth doll named "Rags." The February 1912 issue of *McCall's* (a magazine to which Gruelle had sold artwork) featured a how-to sewing article entitled "Raggedy Dolls for Little Mothers." Gruelle himself claimed that two of his favorite poems by James Whitcomb Riley ("The Raggedy Man" and "Little Orphant Annie") had inspired Raggedy Ann's name.

While striking for its specificity and newness, Gruelle's 1915 patent design for Raggedy Ann nevertheless echoed attributes of characters from popular illustrated children's books of the time—most notably the moon-face, teardrop nose, and floppy stuffed body of W. W. Denslow's and John R. Neill's scarecrows (appearing in several editions of L. Frank Baum's *The Wizard of Oz*), and the button eyes, triangular nose, and wool hair of Florence Upton's Golliwog characters.

(Clockwise from top) W. W. Denslow's Scarecrow (1899), John Neill's Scarecrow (1904), and Florence Upton's Golliwog (1901). (Author's Collection/Helen and Marc Younger)

Gruelle's early cartoons and drawings often incorporated rag dolls, sometimes shown trailing from the hand of a little girl. (This page, top) New York Herald, *April 15, 1914; (bottom) January 7, 1912; (opposite page left)* New York Herald, *June 13, 1915; (top right) the* Cleveland Press, *March 9, 1910; (bottom right)* Grimm's Fairy Tales *(1914).*

3.—When Dolly finished the sign she ran with it to the bird house. "Maybe we will have a beautiful blue bird for a tenant," she told Lou.

6.—What was Dolly's surprise and happiness when she went in and found a dear Mamma Jenny Wren and her children all neatly settled in their new home.
"I hope you will sing often," said Dolly.
"Indeed, I will" said Jenny, "for when I sing I do not have time to think of anything except happiness."

continued from page 35

character dolls), would have been logical purveyors of a Raggedy Ann doll.

However, despite the fact that dozens of new rag dolls—many not as appealing or well-designed as Raggedy Ann—filled the pages of *Toys and Novelties* and *Playthings*, Gruelle's new doll was nowhere to be seen, much less promoted or advertised. It appears that no company was clamoring for the rights to manufacture a Raggedy Ann doll.

Gruelle later confirmed that his own doll making during the summer of 1915 had been modest and short-lived, resulting in dolls that he purveyed on a small scale to protect his Raggedy Ann trademark (which was valid only as long as his dolls continued to be made and sold). Except for these few home-made dolls, there would be no noticeable Raggedy Ann presence in the playthings marketplace for several more years.

Gruelle may have been reluctant (at least initially) to assign manufacturing rights for his new doll, believing that he and his family could sustain their own cottage industry, and thus retain complete control of Raggedy Ann's design and trademark. Other circumstances also undoubtedly affected his doll marketing plans. Mere months after his design patent was granted and only two weeks before his trademark was approved, Gruelle's thirteen-year-old daughter, Marcella, died following a lengthy illness caused by an unsterilized vaccination needle.

Perhaps most relevant to Raggedy Ann's slow start was the fact that as of 1915 she was not a recognizable character. Unlike Rose O'Neill's Kewpies,

You see, Johnny Gruelle, who made them, once lived in Toyland himself. That is why he knows so well just how lively toys ought to look, no matter what they are doing or how they are dressed.

"Who Are Quacky Doodles and Danny Daddles?" 1916

Winsor McCay's Little Nemo, or even Gruelle's own Mr. Twee Deedle, Raggedy Ann did not have a public presence or identifiable persona. Consequently, she lacked the popular recognition that bred familiarity and led to doll sales. Though endearing and embraceable, Raggedy Ann was a virtual unknown.

Gruelle may have known that he needed to give his Raggedy Ann a literary life before expecting a manufacturer's interest. His years of newspaper, magazine, and book work had certainly exposed him to the power of the written word and pictured gesture in building and sustaining a character in the minds of readers, particularly children.

In any event, in 1915 Johnny Gruelle approached the Chicago-based P. F. Volland & Company armed with a set of rag-doll poems he had written a decade earlier, verses he modestly had dubbed "bum verse." On August 30, 1915, Gruelle signed a contract with the company whose specialties were ornate greeting

This agreement superseded by agreement of April 22/18.

MEMORANDUM OF AGREEMENT

made the 30th day of August, 1915, between JOHN B. GRUELLE of Norwalk, Connecticut, and P. F. VOLLAND & COMPANY, a co-partnership consisting of P. F. Volland and F. J. Clampitt, hereinafter referred to as Volland.

WHEREAS Gruelle is the author of and proprietor of copyright in an illustrated book identified herein as the RAGGEDY ANN BOOK; and

WHEREAS Volland is desirous of acquiring the exclusive right to print, publish and sell the said RAGGEDY ANN BOOK;

IN CONSIDERATION of the said exclusive right, Volland agrees to pay to Gruelle on each RAGGEDY ANN BOOK sold a royalty of eight per cent (8%) of the list retail price, all royalties accruing up to January 1 of each year to be due and payable on or before the following 15th day of January, provided however, that on January 1, 1916, Volland shall pay to Gruelle two hundred dollars ($200.00) on account of royalties to accrue. Volland further agrees that Gruelle or the latter's accredited agent may have access to Volland's books or records of account for the purpose of ascertaining the correct figures concerning the printing, publishing and sale of the RAGGEDY ANN BOOKS.

Gruelle agrees not to prepare, for publication by any publisher, person or concern other than Volland, any drawings, posters or books, bearing the name or having the subject matter of the RAGGEDY ANN BOOK.

IT IS FURTHER AGREED by the parties hereto that the exclusive right herein granted may be terminated by Gruelle upon Volland's failure to pay the royalties as herein agreed upon or, in case the accrued royalties for any year beginning January 1, 1918, or thereafter, shall not amount to one hundred dollars ($100.00), such termination of the right being effective upon ninety (90) days' written notice from Gruelle to Volland

John B. Gruelle

P. F. VOLLAND & CO.
By P. F. Volland

The milestone contract that started it all. (Cox-Scheffey Collection)

(21)

Among the Quacky Doodles and Danny Daddles tie-in merchandise was this printed fabric. (Garcia-Garst Booksellers)

The ever-popular Quacky Doodles family appears again in the 89th release of Paramount-Bray-Pictographs to furnish another decidedly funny animated cartoon, and no doubt Paramount audiences will have further occasion for hearty laughter at the antics of these quaintly funny little duckies.

Press release
October 1917

---

(22)

Gruelle modeled his Cleety the Clown character (introduced in Quacky Doodles' and Danny Daddles' Book*) (above) after a Schoenhut clown called "the Card Player"(right). Cleety also appeared in later books, including* Beloved Belindy.

cards and lavishly designed inspirational and juvenile books. In the contract, he agreed to deliver an illustrated children's book identified only generically as "The Raggedy Ann Book."

As providential as it would prove to be, this first book contract (which, interestingly, made no mention of dolls) was eclipsed by another agreement, dated the same day. In this much more detailed document, Gruelle authorized the Volland Company to produce and sell wooden toy ducks that he had patented on the same day as Raggedy Ann. They were characters he had designed and depicted in illustrations in *Quacky Doodles' and Danny Doodles' Book,* a Volland book written by Gruelle's Silvermine neighbor, Rose Strong Hubbell.

In order to have tie-in merchandise ready to introduce along with its book, Volland arranged with the A. Schoenhut Company of Philadelphia to

WE'RE THE HAPPY NIMBLE TOYS: QUACKY DOODLES AND DANNY DADDLES

On August 30, 1915, Johnny Gruelle licensed the P. F. Volland Company to sell wooden toy ducks based on his patented designs. Production did not begin until early February 1916 when Volland began jobbing duck production to A. Schoenhut of Philadelphia—a company well known for its toy pianos, realistic movable-limbed dolls, and wooden character playthings.

The toy ducks were a part of Volland's newly devised scheme to purvey high-quality merchandise as tie-ins to its books. In fact, as dozens of bright yellow ducks were being shipped to Volland, presses in Chicago were already rolling with *Quacky Doodles' and Danny Daddles' Book*, a book of verses written by Rose Strong Hubbell and illustrated by Gruelle.

The ducks formed a fanciful band. The largest duo, Mama Quacky Doodles and Daddy Danny Daddles, measured 10½" and 12¼", respectively. The middle-sized ducks, Miss Quacky Doodles and Danny Daddles Junior, were 8½" and 10". The smallest pair, Baby Quacky Doodles and Baby Danny Daddles, stood 5¾" and 6¼". Mama Quacky, Miss Quacky, and Baby Quacky came with removable cloth bonnets, while their male counterparts were made with attached wooden hats. All were meticulously hand-painted with bills that could be opened slightly and were strung internally, allowing legs and jointed necks to move.

Schoenhut sold the toy ducks to Volland by the dozen for 58 cents, 42 cents, and 22 cents per duck, depending on size. The Quacky Doodles and Danny Daddles toys retailed for $1.50, $1.00, and 50 cents apiece.

Gruelle's royalties for the Quacky Doodles and Danny Daddles toys were set at a modest 2 cents, 1½ cents, and 1 cent per duck. In exchange for Volland's purchasing special advertising in newspapers and magazines, Gruelle had agreed to a maximum of $5,000 in royalties during the first year of production—a moot point, as it turned out, since his first-year earnings amounted to barely half that.

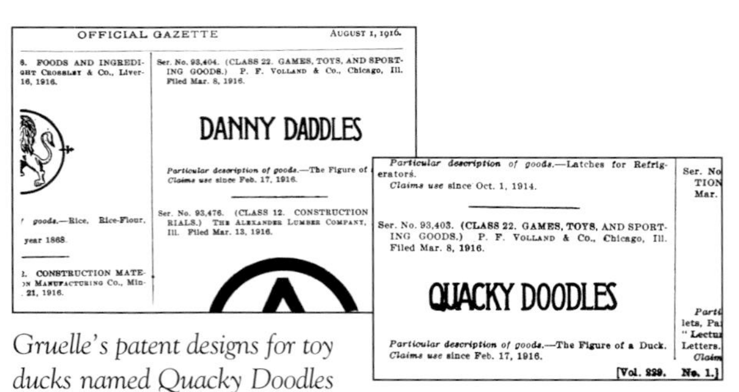

OFFICIAL GAZETTE — AUGUST 1, 1916.

6. FOODS AND INGREDI- ...GHT CROSSLEY & Co., Liver- ...16, 1916.

... goods.—Rice, Rice-Flour, ... year 1868.

2. CONSTRUCTION MATE- ...N MANUFACTURING Co., Min- ...21, 1916.

Ser. No. 93,404. (CLASS 22. GAMES, TOYS, AND SPORTING GOODS.) P. F. VOLLAND & Co., Chicago, Ill. Filed Mar. 8, 1916.

DANNY DADDLES

Particular description of goods.—The Figure of ...
Claims use since Feb. 17, 1916.

Ser. No. 93,476. (CLASS 12. CONSTRUCTION ...RIALS.) THE ALEXANDER LUMBER COMPANY, ... Ill. Filed Mar. 13, 1916.

Particular description of goods.—Latches for Refrigerators.
Claims use since Oct. 1, 1914.

Ser. No. 93,403. (CLASS 22. GAMES, TOYS, AND SPORTING GOODS.) P. F. VOLLAND & Co., Chicago, Ill. Filed Mar. 8, 1916.

QUACKY DOODLES

Particular description of goods.—The Figure of a Duck.
Claims use since Feb. 17, 1916.

[Vol. 229. No. 1.]

Gruelle's patent designs for toy ducks named Quacky Doodles and Danny Daddles were registered on September 7, 1915, for a term of fourteen years; no designs for Baby Danny Daddles and Baby Quacky Doodles were registered. On March 8, 1916, Volland secured a trademark for the name and logo "Quacky Doodles" and "Danny Daddles," which had been in use since February 17.

The bonneted and bespectacled ducks that Gruelle included in early episodes of "Mr. Twee Deedle" were unmistakable Quacky Doodles forebears.

The movable legs and jointed necks of the handpainted Quacky Doodles and Danny Daddles toys were held together by Schoenhut's patented internal stringing.

During 1916 and 1917, Volland received as many as three or four shipments of toy ducks per month. However, when World War I began, many Schoenhut employees returned to Germany, forcing the manufacturer to cancel its Quacky Doodles/Danny Daddles contract. Volland placed its final order for ducks in early 1918 and continued to sell the toys until stock was exhausted.

Although in 1935 Gruelle considered reviving his Quacky Doodles and Danny Daddles toys, his quirky little ducks were never manufactured again. However, the wooden toy ducks later produced by Fisher-Price, named "Dr. Doodle" and Granny Doodle" <1931>, "Pushy Doodle" <1933>, "Ducky Daddles" <1941>, and "Quacky Family" <1941>, left no doubt as to their inspiration.

The only clues that Quacky Doodles and Danny Daddles were manufactured by Schoenhut and designed by Gruelle were dates on a label on each duck's foot: December 2, 1902 (the patent date for Schoenhut's exclusive stringing process), April 14, 1903 (the patent date for Schoenhut's Humpty Dumpty circus), and September 7, 1915 (the date Gruelle's design patents were registered).

THE TALK OF THE TOWN

The Quacky Doodles Family

THE LOVABLE LAUGHABLE NIMBLE TOYS

These friendly, funny, and clever ducks are the world's wisest and funniest toys. They stand up and sit down, they laugh and do tricks—in fact do everything but talk. They are painted by hand.

(24)

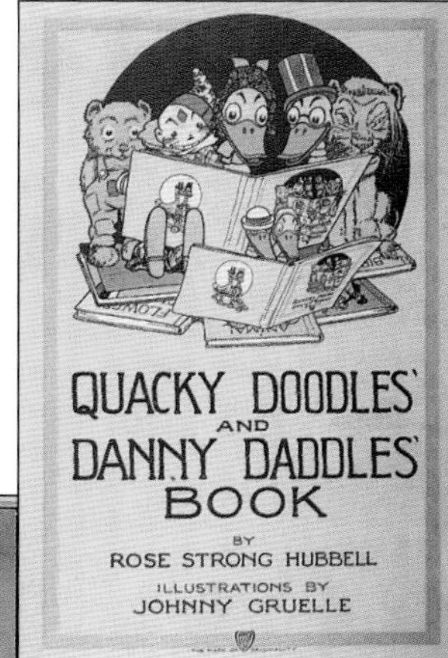

Gruelle's illustrations in Quacky Doodles' and Danny Daddles' Book *(1916) introduced his toy ducks to a young readership.*

(26)

Volland advertised Quacky Doodles and Danny Daddles in space ads, company brochures, and booklets. Stores that carried the toys and books included Best & Company (New York City), Jordan Marsh Company (Boston), and A. E. Little Company (Los Angeles). (Gloria Timmell/The Chicago Public Library)

(25)

Though manufacturing was discontinued in 1918, Quacky Doodles' and Danny Daddles' images appeared throughout the 1920s in advertising for gelatin, floor coverings, and bakery goods.

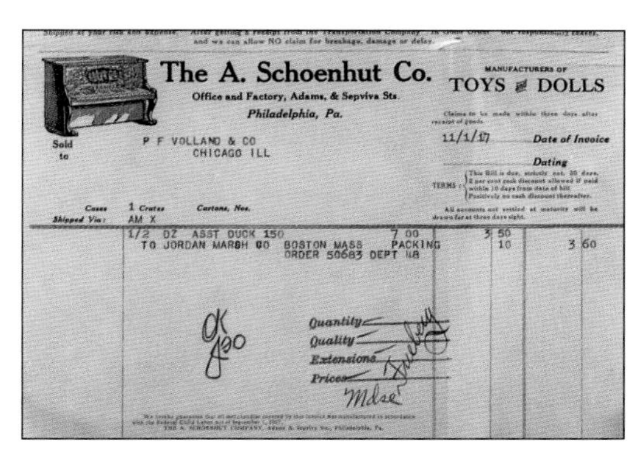
The A. Schoenhut Co.

Office and Factory, Adams, & Sepviva Sts.

Philadelphia, Pa.

MANUFACTURERS OF TOYS & DOLLS

Sold to P F VOLLAND & CO CHICAGO ILL

11/1/17 Date of Invoice

1/2 DZ ASST DUCK 150 TO JORDAN MARSH CO BOSTON MASS ORDER 50683 DEPT 48 7 00 PACKING 3 50 10 3 60

Quantity
Quality
Extensions
Prices

The Schoenhut Company shipped multiple orders of ducks to Volland between 1916 and 1918. (The Chicago Public Library)

EARLY BIRDS:
THE QUACKY DOODLES CARTOONS

In 1916, producer John R. Bray enlisted Johnny Gruelle for a series of animated cartoons featuring Quacky Doodles and Danny Daddles. A pioneer in the still-nascent world of animation and owner of his own animation studios in New York City, Bray was under contract to supply Paramount with 1,000 feet of film footage per week to be shown in local movie theaters.

Billed as "The Magazine of the Screen," these Paramount-Bray "Pictographs" combined educational newsreels, travelogues, and cartoons and were screened alongside feature films. Bray, who was familiar with Gruelle's work (both were cartoonists for *Judge* magazine), had decided that Gruelle's whimsical little ducks would be perfect Pictograph characters.

The resulting Quacky Doodles cartoons (six in all) were simple and slapstickish, relying on the battle of the sexes, prohibition, and wartime patriotism to drive their storylines. Titles and release dates were as follows:

- "The Quacky Doodles' Picnic" (2/17/17)
- "The Quacky Doodles' Food Crisis" (3/6/17)
- "The Early Bird" (3/19/17)
- "Soldiering for Fair" (4/19/17)
- "Quacky Doodles Signs the Pledge" (9/7/17)
- "The Cheater" (10/8/17)

Records indicate that Gruelle had a direct hand in animating the earliest of the six cartoons; a Bray staff animator, F. M. Follett, is credited on several others. A Volland Company ledger indicates some kind of royalty arrangement with Bray studios. However, the absence of entries points to a short, unlucrative life for the productions.

(27)

QUACKY DOODLES HOOKS UP WITH PARAMOUNT.

Patrons of Paramount theaters have a brand new laugh in store for them when they see the Paramount-Bray-Pictographs to be released February 17. On that release will for the first time be shown a group of little characters which have heretofore furnished unending amusement to hundreds of thousands of children and parents from between the covers of books and as life-like toys. This is the Quacky Doodles family, made up of Quacky Doodles, Danny Doodles and the little Doodles.

The "Quacky Doodles" Family.

Johnny B. Gruelle, who attained international fame some few years ago when he won the thousand dollar prize offered by the New York Herald for the best comic, is the creator of these funny little creatures, and he has been engaged by the Bray Studios, Inc., to animate them for Paramount patrons. They will therefore appear at regular intervals, together with Col. Heeza Liar and Bobby Bumps and his famous dog, Fido.

Easily the characters that strike the average person as being funniest are birds and animals humanized. In the pages of comic weeklies and supplements and the Sunday supplements of newspapers we laugh heartiest at the antics and expressions of such humanized animals, and so the Quacky Doodles family have gained their tremendous popularity among children of all ages because they are so truly comical and do such excruciatingly funny things.

Paramount-Bray publicized its Quacky Doodles cartoons with posters and stories in movie magazines such as The Moving Picture World. *Low copyright registration numbers for the cartoons (between #852 and #1043) hint that the Quacky Doodles releases were among the earliest motion pictures ever to be copyrighted.*

begin manufacturing six different Quacky Doodles and Danny Daddles toys to be sold under the Volland label. When they premiered in May 1916, the Quacky Doodles book and toys were promoted with colorful flyers and space advertisements and would even spawn a short-lived series of animated cartoons.

During the next few years, Gruelle continued writing and illustrating books for Volland. Though none focused solely on Raggedy Ann, Gruelle slipped drawings of his new rag-doll character (and the book about her that he was obviously working on) into his illustrations.

Manufacturing Totals
Volland Quacky Doodles and Danny Daddles Toy Ducks

The following figures are based on invoices, ledger entries, and other records. Figures represent combined totals for all three sizes manufactured for the P. F. Volland Company by the A. Schoenhut Company of Philadelphia.

	1916	1917	1918
January	0	16,800	516
February	1,704	780	0
March	2,280	0	0
April	0	4,920	0
May	7,740	0	0
June	0	0	0
July	5,664	0	0
August	17,952	6,144	0
September	6,132	0	0
October	12,300	0	0
November	23,208	96	0
December	23,232	0	0
Annual Totals	**100,212**	**28,740**	**516**

Grand Total 129,468

When wartime downsizing of the Schoenhut Company forced a halt in the production of Quacky Doodles and Danny Daddles toys in early 1918, Volland went in search of another book-toy tie-in. On April 19, the company turned to Gruelle, offering him a bonafide contract for his Raggedy Ann.

In this lengthier document (which supplanted the one executed in 1915), Volland agreed to publish a book written and illustrated by Gruelle specifically titled *Raggedy Ann Stories*, promising Gruelle a 5 percent royalty plus a $200 fee for his illustrations. Unlike the previous contract, this one also contained a clear provision for the Volland Company to sell Raggedy Ann dolls, whose design would be based on Gruelle's *Raggedy Ann Stories* illustrations. For those dolls used solely as retail display props to promote his book, Gruelle would be paid no royalties. However, for those dolls that Volland sold, Gruelle would receive a royalty of 5 percent.

In the fall of 1918, as the 96-page *Raggedy Ann Stories* was readied for publication, the Volland Company placed an order with the Non-Breakable Doll & Toy Company of Muskegon, Michigan, for twenty-four dozen Raggedy Ann dolls. Though this modest first order was probably for dolls intended as retail display props, Volland's subsequent orders (totalling more than 13,000 by the end of 1919) confirm that Raggedy Ann dolls were being actively retailed to a buying public.

The Volland Company was in an excellent position to do well by Raggedy Ann. For one thing, a book-and-doll combination was more marketable than either by itself. For another, the wartime embargo on German goods (including dolls) and the

A hint of another book Gruelle was working on for Volland appeared in his illustrations for Quacky Doodles' and Danny Daddles' Book.

In January 1918, Physical Culture *ran this photograph of illustrator John Gruelle and his young son Worth.*

subsequent 75 percent tariff secured from Congress by the Toy Manufacturers of the United States left American doll and toy purveyors with much less foreign competition. And, though sales for the Quacky Doodles books and toys had been only moderate, figures for Raggedy Ann confirmed that, this time, Volland had hit on a winning combination.

But one thing was certain. Without the literary persona and make-believe adventures created by Gruelle in *Raggedy Ann Stories*, Raggedy Ann—the doll—might never have become known beyond Gruelle's home town of Norwalk, Connecticut.

Not to be confused with the Raggedy Ann dolls produced in 1915 by Gruelle himself to secure his trademark, these first commercial "Cottage" Raggedy Ann dolls, produced for Volland by Non-Breakable, were intentionally patterned after Gruelle's illustrations in *Raggedy Ann Stories*, underscoring the contractual link between doll and book.

On March 12, 1920, Johnny Gruelle signed yet another contract with Volland. This one authorized the company to sell a newly designed Raggedy Ann doll, to be patterned after his 1915 patent, for as long as Volland remained the exclusive publisher of his Raggedy Ann books. That same month, Muskegon Toy & Garment Works began turning out

THE VOLLAND BOOKS

The P. F. Volland Company published twenty-six books written and/or illustrated by Johnny Gruelle. All but two of them were issued in the company's "Happy Children" and "Sunny Book" series.

The 96-page "Happy Children" titles and the 40-page "Sunny Books" are printed on matte coated paper, have illustrated board covers, and were issued in gift boxes resembling shallow stationery boxes. The Volland publisher's device usually appears on the title page, and often (but not always) editions are stated in italics on the copyright page. Prior to 1928, covers of the "Happy Children" and "Sunny Books" had continuous paper labels covering not only front and back covers, but also spines; by 1928, the books' spines were visible cloth.

Although illustrations in pre-1927 titles were in full color, Gruelle's artwork for new Raggedy Ann and Andy titles published after 1927 appeared in both color and black-and-white. These later Volland editions served as the model for M. A. Donohue & Company when it began reprinting Gruelle's books in 1934.

Two other Gruelle books—*The Magical Land of Noom* (1922) and *Raggedy Andy's Number Books* (1924)—were not part of the Volland "Happy Children" or "Sunny Book" series. *Noom* was issued in a box in the "Volland Classic" series, and the linen-like *Number Book*, in Volland's "Cloth Art Book" series.

(28)

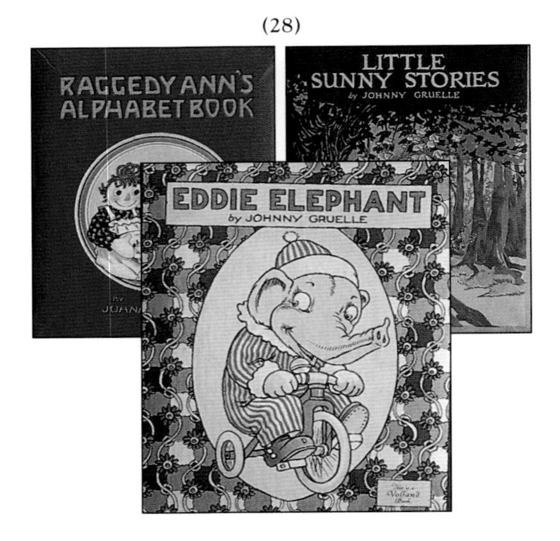

(29)

The Volland Company issued Gruelle's "Happy Children" and "Sunny Books" (shown here) in special gift boxes.

(30)

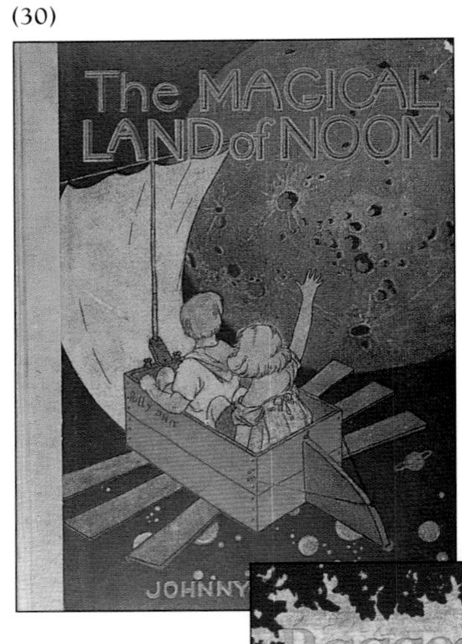

(31)

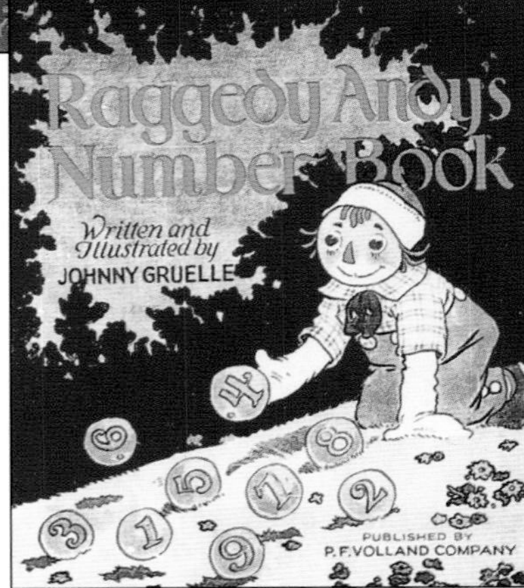

Volland editions of The Magical Land of Noom *(1922) and* Raggedy Andy's Number Book *(1924).*

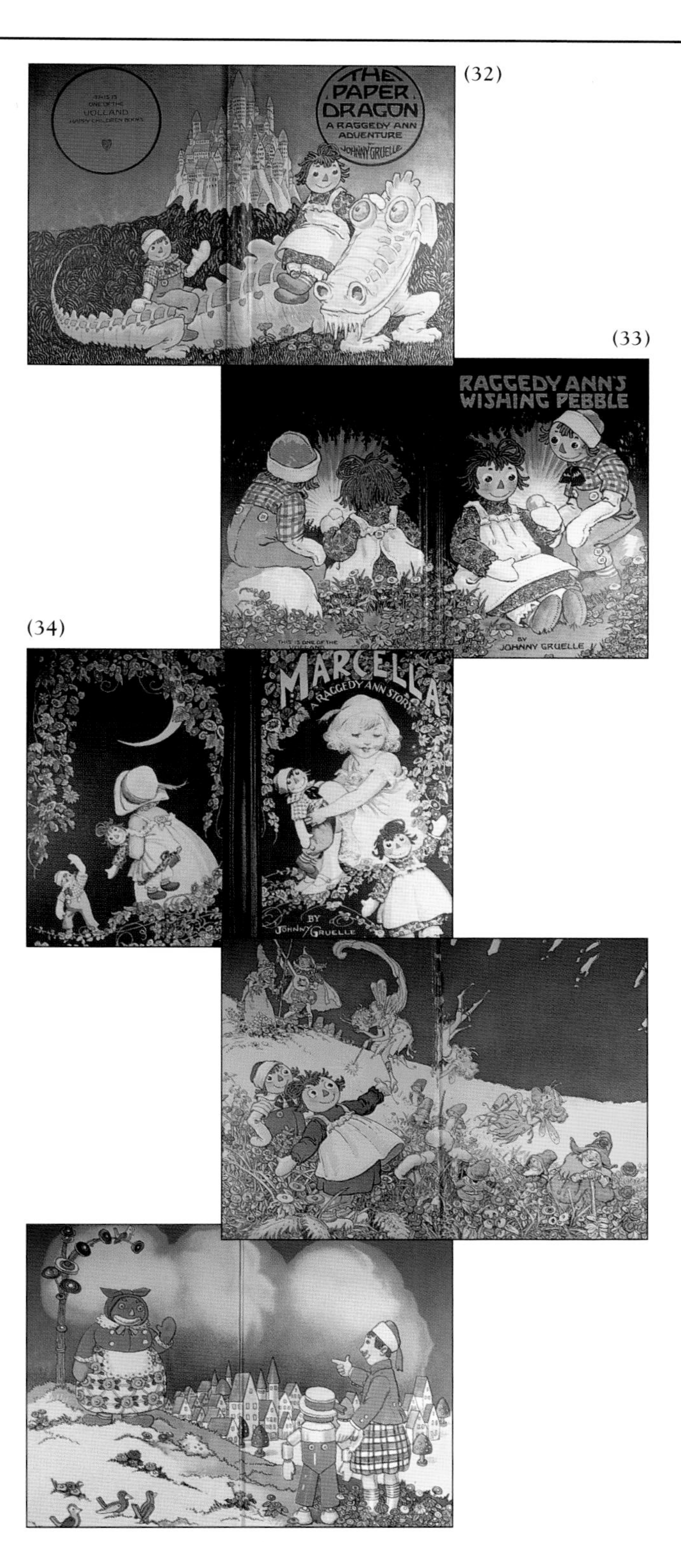

Gruelle's Volland books featured front-and-back cover artwork and lavishly illustrated endpapers.

The Volland Company advertised its offerings with colorful flyers and bookmarks included with its boxed books. (Barbara Lauver/Author's Collection)

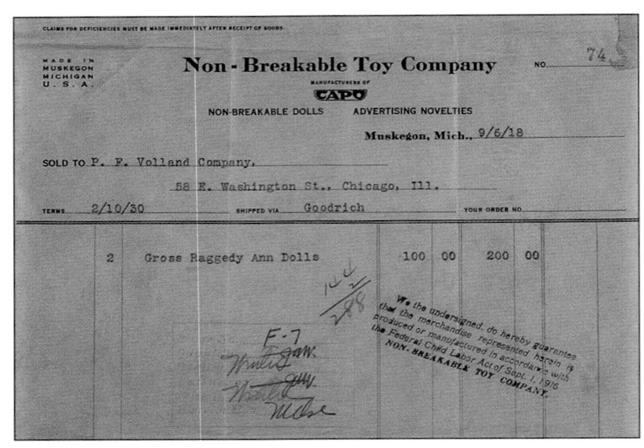

CLAIMS FOR DEFICIENCIES MUST BE MADE IMMEDIATELY AFTER RECEIPT OF GOODS

MADE IN MUSKEGON MICHIGAN U. S. A.

Non - Breakable Toy Company NO. 74

MANUFACTURERS OF CAPO

NON-BREAKABLE DOLLS ADVERTISING NOVELTIES

Muskegon, Mich., 9/6/18

SOLD TO P. F. Volland Company,

58 E. Washington St., Chicago, Ill.

TERMS 2/10/30 SHIPPED VIA Goodrich YOUR ORDER NO.

2	Gross Raggedy Ann Dolls	100 00	200 00	

F-7

We the undersigned, do hereby guarantee that the merchandise represented herein is produced or manufactured in accordance with the Federal Child Labor Act of Sept. 1, 1916
NON-BREAKABLE TOY COMPANY.

The earliest Raggedy Ann dolls, manufactured for Volland by the Non-Breakable Doll & Toy Manufacturing Company, were intended both for display and retail sale. (The Chicago Public Library)

(39)

The "Cottage" Raggedy Anns, produced by Non-Breakable Toy Company for Volland (above), closely resembled Gruelle's art-work for Raggedy Ann Stories *(below). (Photographs by W. Jackson Goff)*

(40)

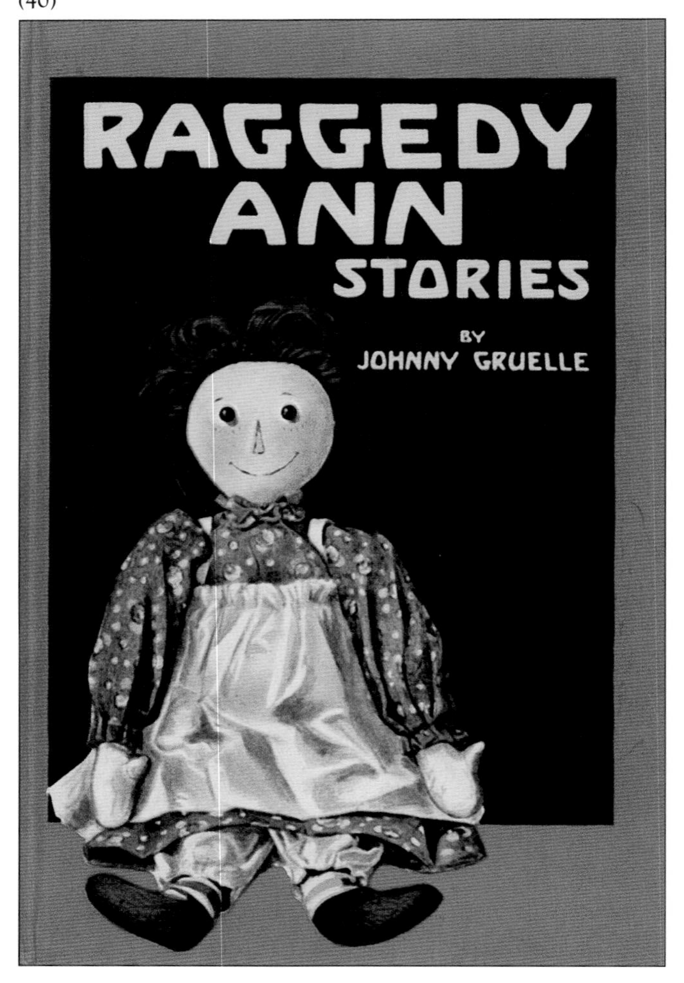

THE TRAGEDY OF THE MINIATURE

Johnny Gruelle's Raggedy Ann had been on the market a scant six months when a woman named Vera Trapagnier strode into the Chicago office of Paul Volland and shot the 45-year-old company president through the heart with a small pistol. At issue was a seemingly benign heirloom—a framed miniature of Gen. George Washington painted by famed Colonial artist John Trumbull.

Trapagnier had acquired the piece as a gift, and Paul Volland had agreed to reproduce it for sale. Though the Volland Company maintained that fewer than 200 reproductions had been sold and that Trapagnier had received a $500 advance, Trapagnier claimed she was never paid any royalties and that Volland would not surrender the original until she paid the $174 it had cost to reproduce it.

A defense fund for the 60-year-old Trapagnier was established by several artists who felt they also had been victims of Paul Volland's "business habits." But in August 1919, Trapagnier was convicted of manslaughter and sentenced to an unspecified term to be served at the Illinois State Penitentiary.

For Gruelle, who was just beginning to feel established at the Volland Company, news of the slaying not only highlighted the tensions between Volland and those he did business with; it also set in motion changes and retrenchments that would forever affect the future of Raggedy Ann and Andy.

CHIEF CHARACTERS IN "THE TRAGEDY OF THE MINIATURE"

Slain Publisher, His Divorced Wife, and the Woman Who Killed Him, Photographed on Way to Jail, and the Painting of George Washington, Which Led to the Murder.

MRS. VERA TREPAGNIER, who killed the publisher.

MINIATURE OF WASHINGTON MADE HER KILL

P. F. Volland Slain in Office by Widow 60 Years Old.

(Continued from first page.)

A copy of the John Trumbull miniature of George Washington and the walnut frame, in which it has been encased for a hundred years or more.

PAUL FREDERICK VOLLAND, the slain man (above) and MRS. VOLLAND, who divorced him some time ago.

Member of Several Clubs.

A full-page article detailing the grisly Paul Volland slaying appeared in the May 6, 1919, issue of the Chicago Daily Tribune. *(Lauren Bufferd)*

more mass-produced-looking "Patent" Raggedy Ann dolls, each of which was clearly stamped on its upper torso with the line "Patented September 7, 1915."

With his Raggedy Ann dolls and books fast becoming fixtures in playrooms across the country, what should have been a heady, exhilarating time for Johnny Gruelle was blighted in the spring of 1919 when he received word that Paul Volland had been gunned down in his Chicago office. Volland's parsimony notwithstanding, Gruelle had looked upon the company president as a visionary whose perfectionism and marketing acumen had virtually insured Raggedy Ann's success. Johnny Gruelle was now left wondering about his future.

Subdued but intent on not losing momentum, Gruelle wrote immediately to new company president F. J. Clampitt, securing a renewed commitment to the boy rag doll that Paul Volland had agreed to a short time before he died. Gruelle set to work on a new book and doll design, and on March 16, 1920, submitted again to the U.S. Patent Office. On August 24, a patent was registered in his name for a period of fourteen years. Though generically drawn, undressed, and unnamed, the design (not to mention the timing of the submission) left little doubt that this patent was intended to protect Gruelle's newest rag-doll creation, an impish scamp he would name Raggedy Andy.

Behind the clouds of worry there is always a rainbow of happiness, if we only knew how to find it as the Dwarfies do.

Johnny Gruelle
1921

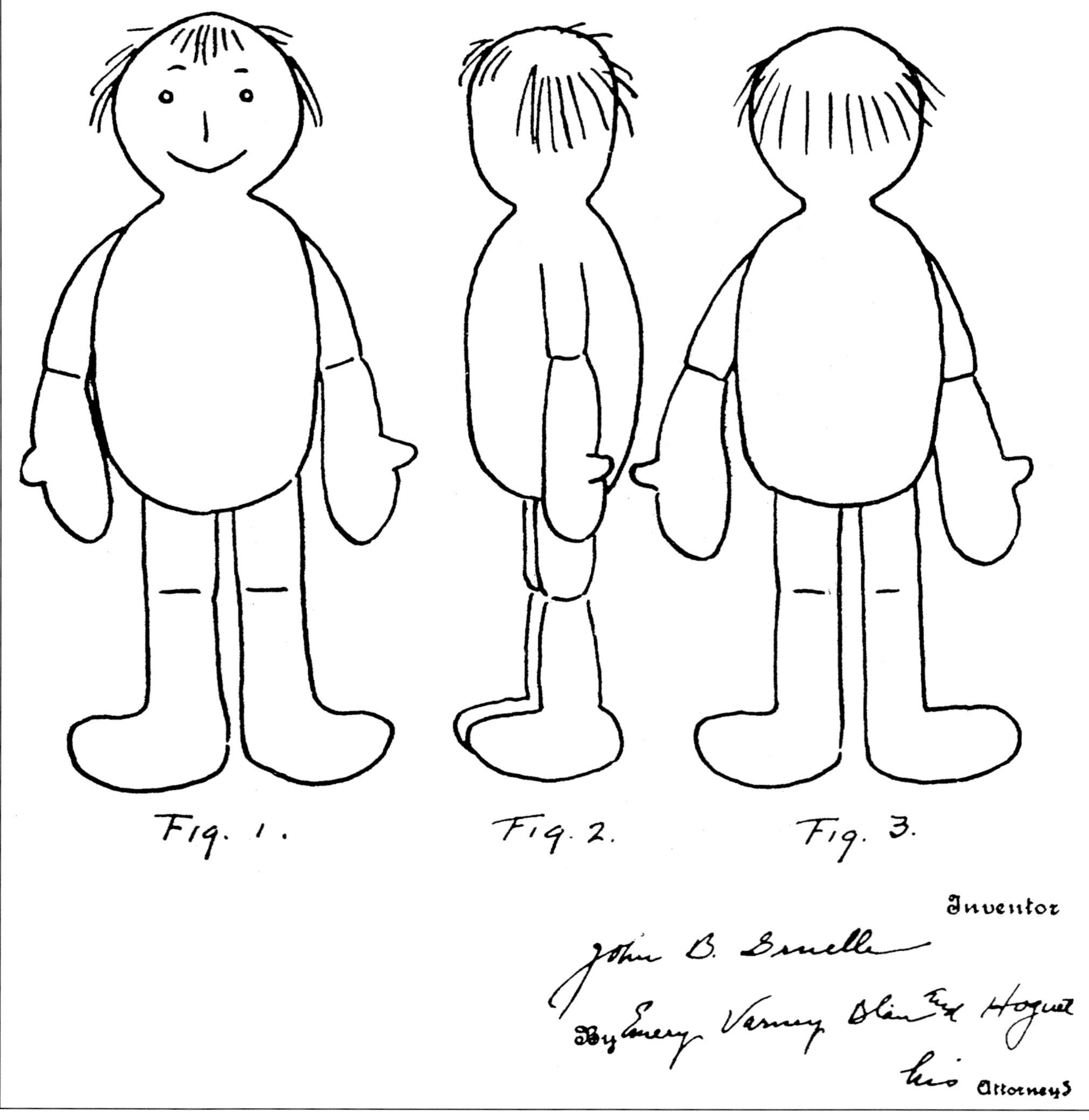

In contrast to the detailed Raggedy Ann design, Gruelle's simplified Raggedy Andy design offered broader patent protection, covering variations in materials, facial and body design, and clothing styles.

INVOICE

The Beers-Keeler-Bowman Co., Inc.

NORWALK, CONN.

Order No.

Invoice No.

Sold To P. F. Volland Co.
Chicago Ill.

Address

Date June 27 '20

Ship to

Address

CUSTOMER'S ORDER NO. ORDER RECEIVED SALESMAN

Terms SHIP VIA

NO. ORDERED	NO. SHIPPED	DESCRIPTION	PRICE	AMOUNT	TOTAL
	180	Andy Dolls	90	162.00	

Delivery not shown by receiving slips. B.K. B. Co. proved delivery by express receipts, and we show evidence of express charges on those dates.

Order No. 9018
Quantity
Quality
Extensions
Prices
Distribution Mdse.

Invoice for Raggedy Andy dolls, Beers-Keeler-Bowman-Company, 1920. (The Chicago Public Library)

Raggedy Andy typifies kindliness, generosity, sympathy and fun; a happy little creature stuffed with sunshine and wearing the badge of pleasantry, a broad perpetual smiley smile and a look of honesty in his shoe button eyes.

Johnny Gruelle
1920

Gruelle's portrayals of brother-and-sister Dickie and Dolly in "Mr. Twee Deedle" had proven how engaging a brother-and-sister team could be. So when he brought Raggedy Andy to life in *Raggedy Andy Stories*, Gruelle knew to cast him as Raggedy Ann's long-lost brother and to engage the dolls in a hearty share of sibling-style banter and adventure.

Though the Volland Company would be the exclusive purveyor of Raggedy Andy, when it came to doll production, Gruelle insisted on selecting and dealing directly with a manufacturer. The fledgling company he chose, Beers-Keeler-Bowman, just happened to be located in his home town of Norwalk, Connecticut. In fact, Gruelle may even have helped persuade his close neighbor, former stationer William Beers, to launch his doll manufacturing business, which opened its doors in 1920. In any event, Raggedy Andy was among the first dolls Beers-Keeler-Bowman produced.

Though Beers-Keeler-Bowman would eventually relocate to larger quarters, its first "factories" were located downtown, in three-story, walk-up brick buildings already outfitted for sewing and assembling the apparel and millinery items for which Norwalk was famous. Here were made the first two dozen Raggedy Andy samples, dolls that members of the Gruelle family reputedly had a hand in fabricating or hand finishing.

continued on page 64

THEME AND VARIATIONS: THE VOLLAND RAGGEDY ANN AND RAGGEDY ANDY DOLLS

The thousands of Raggedy Ann and Andy dolls sold by P. F. Volland are more notable for their differences than for their similarities. The fact that Volland jobbed doll production simultaneously to several different manufacturers accounts for the more obvious visual diversity. Subtler variations among dolls are due to, among other things, availability or shortages of materials, differing equipment and procedures, shifting personnel, varying workmanship, economization, and sometimes, to whim alone.

Identifying and sequencing the Volland Raggedy Ann and Andy dolls is a complex and somewhat speculative task, further complicated by the fact that most dolls were not marked, and few were photographed as part of company record-keeping. In categorizing the Volland (or any other) dolls and making sense of their design differences, the collector is cautioned against assuming motives or intention on the part of designer or company where none may have existed. With this in mind, and using what reliable documentation is available, the Volland Raggedy Ann and Andy dolls can be placed in present-day classifications organized chronologically.

(Note: Doll identifiers that appear in quotation marks are either commonly used terms or the author's nomenclature—not official company designations. They are presented here to assist in identifying and classifying representative Volland dolls.)

"Cottage" Raggedy Anns (1918-19). The earliest Raggedy Ann dolls offered by the P. F. Volland Company were manufactured by the Muskegon, Michigan-based Non-Breakable Doll & Toy Company. Though machine sewn, these Raggedy Anns (designated here as "Cottage" dolls) wear the hallmarks of by-hand, "cottage industry"-style fabrication. Some believe that Johnny Gruelle may have had a hand in fabricating these "Cottage" dolls—an apparently apocryphal account that is difficult to confirm or refute.

Volland's introduction of the "Cottage" dolls (which were modeled after Gruelle's book illustrations) was timed to coincide with the fall 1918 publication of *Raggedy Ann Stories*. Though some of the "Cottage" dolls were undoubtedly used as retail display props, most were manufactured to be sold as related merchandise along with Gruelle's new book.

(41)

"Cottage" Raggedy Anns (1918-19). *Records confirm that approximately 13,000 "Cottage" Raggedy Ann dolls, patterned after illustrations for* Raggedy Ann Stories, *were produced by Non-Breakable Doll & Toy Manufacturing Company. The dolls sold to Volland for 70 (later 75) cents each.*

The earliest "Cottage" Raggedy Ann dolls stand approximately 15" tall and have dainty muslin bodies stuffed with cotton batting and wigs of medium-weight brown wool yarn. Distinguishing features include hand-applied faces (a painted-on line smile, a narrow, red triangular nose thinly outlined in black, eyebrows and lower lashes, sewn-on black shoe button eyes), painted-on Mary Jane-style shoes (through which the red-and-white-striped leg fabric was visible), and the distinctive blue floral print of the dolls' challis dresses. This was a striking match for the dress fabric Gruelle had depicted on the cover of *Raggedy Ann Stories*. These Raggedy Anns also wear white cotton pinafores and pantalettes, and each has a small, dense cardboard heart set between cotton stuffing and chest fabric—a practical alternative to the candy heart Gruelle had given Raggedy Ann in his book.

Some time in 1919, Non-Breakable began turning out larger "Cottage" Raggedy Ann dolls. Like the earlier versions, these dolls (some of which measure as tall as 18") have cotton-stuffed muslin bodies, hand-applied faces (some of which incorporate prominent lower lashes and narrow crescent-shaped smiles), shoe- button eyes, cardboard hearts, and wigs of medium-to-dark-brown wool yarn. However, in contrast to the earlier dolls, the later "Cottage" dolls have sewn-on (rather than painted) black or brown fabric feet and wear challis frocks of earth-tone (rather than the original blue-and-white) prints along with their white cotton pantalettes and pinafores. In some cases they possess blue-and-white- (rather than red-and-white-) striped legs.

"Patent" Raggedy Anns (1920-26). In the spring of 1920, Johnny Gruelle authorized the Volland Company to begin selling a brand-new Raggedy Ann doll. Unlike the earlier "Cottage" dolls, which were patterned after Gruelle's book illustrations, these Raggedy Anns were contractually linked to Gruelle's 1915 patented design (though, as it turned out, they would not really resemble his drawing). To produce this new doll, Volland turned to Muskegon Toy and Garment Works of Muskegon, Michigan, which stamped each doll with the line "Patented September 7, 1915."

In addition to their distinctive date stamp (which appears on either the back or front upper torso), the "Patent" Raggedy Ann dolls are notable for their machine-applied, rather than hand-painted, faces. This new facial design consists of a fuller triangular nose outlined in black, black-line smile, blushed cheeks, and eyebrows that curve gently downward at the outside, resulting in a gentle, wistful expression.

Much more uniform in design than the "Cottage" dolls, the "Patent" Raggedy Anns measure between 15-16", and have red-and-white-striped legs, cardboard hearts, and black sewn-on feet. The dolls are dressed in removable cotton or challis dresses (usually a paisley or floral print) and white cotton pinafores and pantalettes. Wigs are of heavier brown yarn than the "Cottage" dolls and usually incorporate looped-yarn bangs and a topknot.

During six years of production, the "Patent" Raggedy Anns' design evolved slightly—heads grew rounder and fuller, printed facial features became darker and more prominent, and wigging became lusher. But with their distinctive wistful expressions and date stamps, the "Patent" Anns comprise a readily identifiable grouping. And the fact that extant examples of "Patent" Raggedy Anns far outnumber other Volland doll styles corresponds to the impressive Muskegon Company's manufacturing total of 76,000 Raggedy Ann dolls.

continued on next page

(42)

(43)

(44)

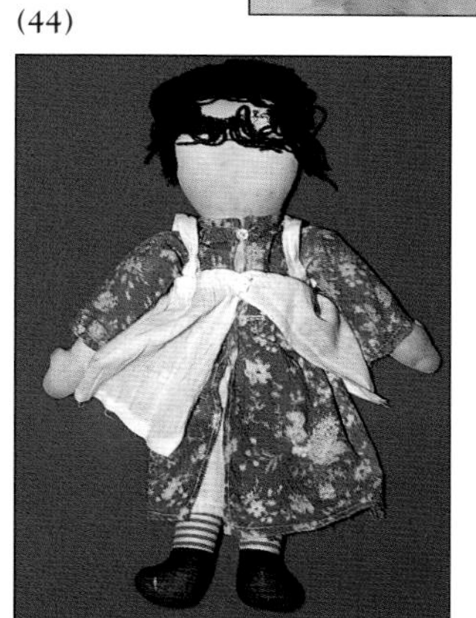

(45)

"Patent" Raggedy Anns (1920-26). *(Bottom row) The date-stamped "Patent" Raggedy Anns were sold to Volland by Muskegon Toy and Garment Works for 80 (later 70, then 60) cents each. (Top) Gerlach-Barklow calendar artist Zula Kenyon incorporated a "Patent" Raggedy Ann in her print "A World of Happiness." (Author's Collection/Candy Brainard/Judy and James Eddy Hatch)*

"Crescent Smile" Raggedy Andys (1920-27). The Norwalk, Connecticut-based Beers-Keeler-Bowman Company began producing Raggedy Andy dolls in June 1920, several months in advance of the planned fall release of Gruelle's *Raggedy Andy Stories*. These robustly styled dolls stand approximately 16" tall and are notable for their substantial heads, disproportionately large hands and thumbs, and most of all, their distinctive, crescent-shaped smiles with red rectangular centers—hence, the designation "Crescent Smile."

Like the "Patent" Raggedy Anns, the "Crescent Smile" Raggedy Andys have muslin bodies, machine-printed faces with shoe-button eyes, red-and-white-striped legs, and sewn-on fabric feet that were cut lower than Ann's. However, in contrast to the "Patent" Anns (which are straight-limbed), the "Crescent Smile" Andys possess stitched-down "joints" on their legs and arms and have no black nose outline. And none of them (nor any subsequent Raggedy Andys) were manufactured with a cardboard heart or with any kind of patent stamp.

The "Crescent Smile" Raggedy Andys' sewn-on clothing consists of a plaid shirt attached to blue cotton pants, set off by mother-of-pearl shirt buttons at the waist and on each trouser leg. The dolls are wigged in various shades of reddish brown yarn, which peeked out of a crescent-shaped blue-and-white cap, tacked on at a jaunty angle.

After buying out his two partners, Keeler and Bowman, in August 1924, William P. Beers became sole owner of the company that would continue, until 1927, as the exclusive manufacturer of the "Crescent Smile" Raggedy Andy. Though the shift in company ownership did not immediately impact the traditional "Crescent Smile" design of the Raggedy Andy dolls, a ten-cent drop in wholesale price (from ninety to eighty cents per doll) in February 1925 may have coincided with the appearance of dolls with scaled-down heads and slightly modified faces.

"Crescent Smile" Raggedy Andys (1920-1927). *Approximately 60,300 "Crescent Smile" Raggedy Andys were manufactured by Beers-Keeler-Bowman/William P. Beers Company during a seven-year production period. A "Crescent Smile" Raggedy Andy was depicted on the January 1924 cover of* Woman's World *and in a 1920s flyer for Mapl-Flake Cereal. (Candy Brainard/Barbara Lauver/Author's Collection)*

"Single Eyelash" Raggedy Anns and Raggedy Andys (mid-to-late 1920s). During the mid-to-late 1920s, a number of Volland Raggedy Ann and Raggedy Andy dolls appeared with single lower eyelashes instead of the customary multiple lashes (usually five for Ann and four for Andy). These distinctive solo facial markings vary, appearing as lines, dots, or inverted triangles or "commas."

Small though it was, this subtle design shift resulted in dolls that better matched the ones Gruelle was depicting in his book illustrations. For though he would continue to draw his dolls with any number of eyelashes (and sometimes with none at all), by the mid-1920s Gruelle had become partial to the more streamlined look of one rather than several lower lashes.

The "Single Eyelash" dolls, measuring between 14-16", bear no company markings or other identifiers and in many respects are similar to their predecessors. However, most "Single Eyelash" Raggedy Andys possess a linear (rather than crescent-shaped) mouth with a small red center, one that more closely resembles Raggedy Ann's.

The intriguing variations among the faces of the "Single Eyelash" dolls suggest that they may have been produced by more than one manufacturer. In fact, the appearance of the "Single Eyelash" dolls coincides with the period when the Beers Company had ceased its manufacture of the standard-size "Crescent Smile" Raggedy Andy, and Volland was placing orders for both Raggedy Ann and Raggedy Andy dolls from C. B. Moore and Gerlach-Barlow.

continued on next page

(52)

(53)

(55)

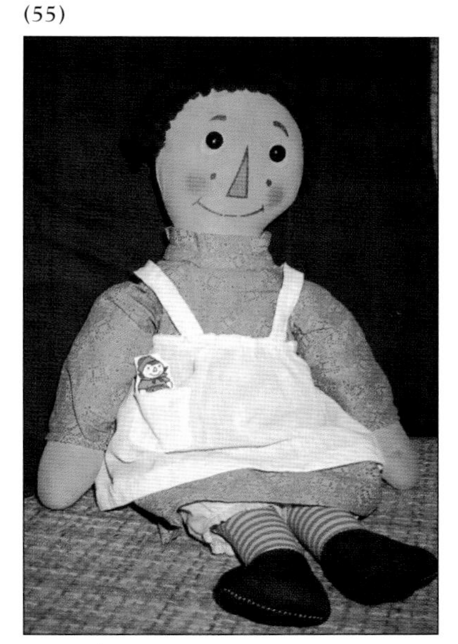

(54)

(56)

"Single-Eyelash" Raggedy Anns and Raggedy Andys (Mid-to-late 1920s). *The variously designed "Single Eyelash" Raggedy Anns and Andy dolls evoked Gruelle's book illustrations (lower left) depicting his rag dolls with only one lower lash. (Kim Avery/Barbara Lauwer/Candy Brainard/Author's Collection)*

"Transitional" Raggedy Anns and Raggedy Andys (late 1920s-early 1930s). During the late 1920s, a 15-16" "Transitional" Volland Raggedy Ann doll appeared. It had an unoutlined nose, heavily blushed cheeks, a low-set smile, and five distinctive triangular lower lashes. At about the same time, a similarly sized but very different looking "Transitional" Raggedy Andy doll, possessing a linear smile and a prominent (sometimes lopsided), unoutlined triangular nose, also found its way into the marketplace.

(57)

(58)

"Transitional" Raggedy Anns and Raggedy Andys (late 1920s-early 1930s). *"Transitional" Raggedy Ann and "Transitional" Raggedy Andy dolls (above). At left, Johnny Gruelle meets a young fan, costumed to match her Volland "Transitional" Raggedy Ann doll. (Candy Brainard/Cox-Scheffey Collection)*

Although these dolls share many attributes with their respective Ann-and-Andy predecessors, their sleeker appearance marked another important step in the evolution toward a more mass- produced-looking doll.

Sales records confirm that the "Transitional" Raggedy Anns and Raggedy Andys would have been manufactured either by C. B. Moore or Volland's parent company, Gerlach-Barklow.

"Finale" Raggedy Anns and Raggedy Andys (1931-33). By 1931, C. B. Moore Company was the only manufacturer supplying Volland with Raggedy Anns and Andys. These last Volland dolls (referred to here as "Finale" dolls) are slightly smaller, standing 14-15" tall. The facial designs, striking for their crispness and uniformity, include heavy eyebrows that resemble bold-face commas. The similarly designed "Finale" dolls were the first-ever Raggedy Ann and Andy to be produced as a matched set.

Marketed as part of an enlarged line of stuffed fabric characters that Volland dubbed the Raggedy Dolls, the "Finale" Raggedy Ann and Andys were economically priced for Depression consumers at $1.50 each.

(59)

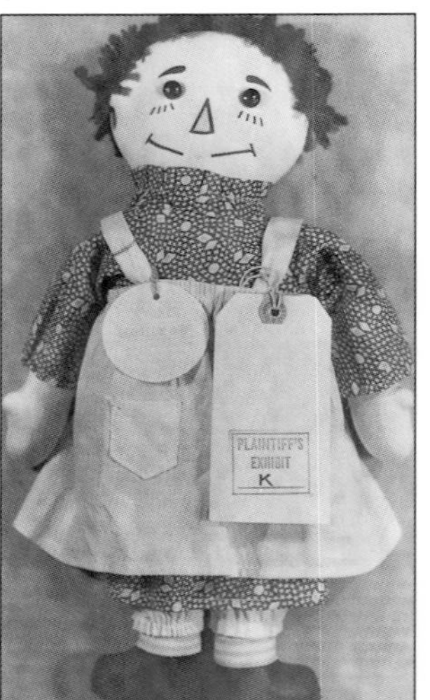

(60)

"Finale" Raggedy Anns and Raggedy Andys (1931-33). *The neatly matched "Finale" Raggedy Ann and Andy dolls (opposite page) were analogous in body and facial design, except that Raggedy Andy lacked Ann's cardboard heart, usually had no outline around his nose, and possessed lower-cut feet. Gruelle's attorneys scrambled to find this set of "Finale" dolls (above) to submit as plaintiff's exhibits in the* Gruelle v. Goldman *lawsuit. Note Andy's gingham, rather than the customary plaid, shirt. (Cox-Scheffey Collection/Barbara Lauver/National Archives)*

"Oversized" Raggedy Anns and Raggedy Andys (1920-21, 1924-25, 1928). Throughout the 1920s, Beers-Keeler-Bowman (later the William Beers Company) produced a limited number of "Oversized" Raggedy Ann and Andy dolls. The first "Oversized" dolls (measuring approximately three feet, give or take a few inches) were not Raggedy Anns but rather Raggedy Andys. One hundred dolls were produced in 1920 and 1921 to be used as retail display props for the newly published *Raggedy Andy Stories* and standard-size Raggedy Andy dolls.

(61)

(62)

(63)

Three years later, in October 1924, the newly consolidated William P. Beers Company again manufactured "Oversized" dolls, this time producing Raggedy Anns as well as Raggedy Andys. These larger dolls possess key characteristics of the separately jobbed standard-sized dolls being produced at the time—Andy, with his familiar crescent-shaped smile, and Ann, with her line smile and proportionately larger cardboard heart.

Both have red-orange noses (Ann's is outlined in black), large, black bead eyes, sewn-on boot-like fabric feet, and the large hands and thumbs that seemed to be a Beers hallmark. Judging from the modest orders placed by Volland (120 Anns and 110 Andys wholesaling to Volland at $4.50 each), these "Oversized" dolls (like those produced in 1920-21) were probably shipped to stores as display items and not retailed to the public.

In December 1928, Beers (which by this time, had ceased production of its standard-size Raggedy Andys) filled an order from Volland for 500 "Oversized" Raggedy Anns and 500 "Oversized" Raggedy Andys, at a wholesale cost of $4.50 per doll. Though no doubt used as retail display props, these "Oversized" dolls (including an "Oversized" Beloved Belindy) were also offered for sale to the general public. Volland's catalogues and back-page book ads touted these dolls as being 30", rather than 36". However, the steep, fifteen-dollar asking price would have been a sizeable outlay for the average individual consumer during the Depression.

"Oversized" Raggedy Anns and Andys (1920-1921; 1924-1925; 1928). *Gruelle posed for a portrait with the first group of "Oversized" Raggedy Andy dolls, produced by Beers-Keeler-Bowman as retail display props and purchased by Volland for $6 each. Later "Oversized" dolls (both Anns and Andys of varying designs) cost Volland $4.50 each. (Norman Meek)*

continued from page 57

Volland placed regular orders with Beers-Keeler-Bowman for dozens of 16-inch "Crescent Smile" (and a small number of "Oversized") Raggedy Andy dolls. When Volland published *Raggedy Andy Stories* on September 17, 1920, retailers all over the country premiered Gruelle's book along with the brand-new dolls.

Rather than compete with Raggedy Ann for a sales niche, the Raggedy Andy dolls and books fueled a resurgence in Raggedy Ann's popularity, proving that a brother-and-sister team was, indeed, more marketable than a single character. Inspired partly by patriotic insistence on non-imported toys and dolls, but mostly by simple fascination with these two new rag dolls, Americans lined up to buy Raggedy Ann and Andy and the books that told their stories.

Volland promoted the Raggedy characters with space advertisements, posters, and retail display cutouts featuring Gruelle artwork. Soon, the company would begin introducing Raggedy-themed specialty merchandise, including valentines and decorated wooden bookshelves (followed later by metal bookends and lithographed picture puzzles). The increasingly popular Raggedys would eventually become Volland Company ambassadors, appearing as anchor characters in the company's general advertising and even making cameo appearances in other Volland authors' books.

Considering the Raggedy's commercial success and the overall growth of the playthings industry during the 1920s, it is puzzling that Volland and

To tell you about Raggedy Ann and Raggedy Andy would be as useless as to tell you about Santa Claus, wouldn't it? Everybody knows them already.
1920s Volland Company catalogue

(64)

The rambunctious Raggedy Andy was a perfect character for Gruelle's little boy readers.

The Beloved

RAGGEDY ANN AND RAGGEDY ANDY DOLLS

RAGGEDY ANN and RAGGEDY ANDY had such good fun in the books Johnny Gruelle wrote:

RAGGEDY ANN AND ANDY
AND THE CAMEL WITH THE WRINKLED KNEES
RAGGEDY ANDY'S NUMBER BOOK
RAGGEDY ANN
RAGGEDY ANDY

that they decided to jump out the books for a little while and be their own lovable, kindly selves.

These friendly rag dolls are as clean and smiling as the many, many colored pictures you see of them in the Raggedy Ann Books. They are well made and stuffed substantially with clean white cotton. Boxed. Price $2.50, each.

P. F. VOLLAND COMPANY

Publishers of Books Good for Children

P. F. Volland Company flyer, early 1920s.

Manufacturing Totals
Volland Raggedy Ann and Raggedy Andy Dolls

The following figures are based on invoices, ledger entries, and correspondence. Figures are rounded to the nearest 100 for standard-sized dolls, and to the nearest 10 for oversized dolls.

Raggedy Ann Dolls

Year	Number Produced		Manufacturer
1918	4,400	standard size	Non-Breakable
1919	8,900	standard size	Non-Breakable
1920	15,600	standard size	Muskegon
1921	21,800	standard size	Muskegon
1922	6,500	standard size	Muskegon
1923	12,400	standard size	Muskegon
1924	6,200 40 40	standard size oversized oversized	Muskegon Beers-Keeler-Bowman William P. Beers
1925	5,500 40	standard size* oversized*	Muskegon William P. Beers
1926	8,000	standard size	Muskegon
1927	7,500	standard size	William P. Beers
1928	6,000 6,000 500	standard size standard size oversized	C. B. Moore Gerlach-Barklow William P. Beers
1929	12,000	standard size	Gerlach-Barklow
1930	6,000 1,500	standard size standard size	C. B. Moore Gerlach-Barklow
1931	6,000	standard size	C. B. Moore
1932	no dolls manufactured		
1933	500	standard size	C. B. Moore

Manfacturer	Standard Size	Oversized	Manufacturer Totals
Non-Breakable	13,300	0	13,300
Muskegon	76,000	0	76,000
B-K-B	0	40	40
Beers	7,500	580	8,080
C.B. Moore	18,500	0	18,500
Gerlach-Barklow	19,500	0	19,500
Grand Totals	**134,800**	**620**	**135,420**

*1925 figures reflect manufacturing totals for January-April only; records for May-December were not available.

Raggedy Andy Dolls

Year	Number Produced		Manufacturer
1920	12,800	standard size	Beers-Keeler-Bowman
	50	oversized	Beers-Keeler-Bowman
1921	12,600	standard size	Beers-Keeler-Bowman
	50	oversized	Beers-Keeler-Bowman
1922	10,600	standard size	Beers-Keeler-Bowman
1923	7,500	standard size	Beers-Keeler-Bowman
1924	1,200	standard size	Beers-Keeler-Bowman
	20	oversized	Beers-Keeler-Bowman
	2,800	standard size	William P. Beers
	50	oversized	William P. Beers
1925	1,800	standard size*	William P. Beers
	40	oversized *	William P. Beers
1926	6,000	standard size	William P. Beers
1927	5,000	standard size	William P. Beers
1928	3,000	standard size	Gerlach-Barklow
	5,000	standard size	C. B. Moore
	500	oversized	William P. Beers
1929	5,000	standard size	Gerlach-Barklow
1930	no dolls manufactured		
1931	8,000	standard size	C. B. Moore
1932	no dolls manufactured		
1933	300	standard size	C. B. Moore

Manfacturer	Standard Size	Oversized	Manufacturer Totals
B-K-B	44,700	120	44,820
Beers	15,600	590	16,190
C.B. Moore	13,300	0	13,300
Gerlach-Barklow	8,000	0	8,000
Grand Totals	**81,600**	**710**	**82,310**

* 1925 figures reflect manufacturing totals for January-April only; records for May-December were not available.

Gruelle were not more aggressive with character merchandising and the licensing of Raggedy Ann and Andy to other companies. Although as early as 1920 Gruelle had considered personally authorizing a trademarked line of Raggedy Ann aprons and dresses for little girls, as well as pursuing other merchandise lines, nothing came of it. Gruelle may have become too tied up with his growing number of magazine and book commissions, and Volland, with its other lines of books, calendars, and greeting cards, for either to take full financial advantage of the Raggedys' ever-increasing renown.

During the next few years, Volland sold an impressive number of Gruelle's dolls and books (which now included several more "Sunny Book" titles and an adventure story for older children, *The Magical Land of Noom*)—all fancifully written and illustrated in his inimitable style. Augmenting

Gruelle's never-published cutout book, The Scissors Mother Goose, *contained pencil sketches of paper dolls and nursery-rhyme first lines interspersed with pages of full-color illustrations. (Joyce Link)*

Every little tot who has followed the adventures of Johnny Mouse and the Woozgoozle each month in Woman's World will, of course, want one of these comfortable Johnny Mouse Jumpers for the warm summer days.

Woman's World
June 1921

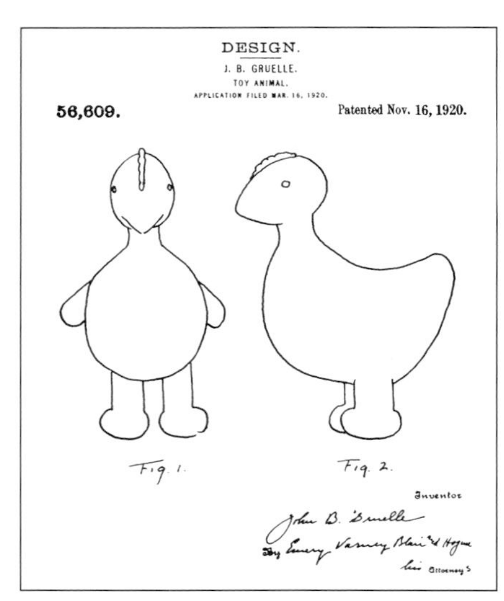

Although plans for Beers-Keeler-Bowman to manufacture a Henrietta Hennypenny character doll based on Gruelle's 1920 patented design never panned out, Gruelle's hen character would appear later in books, including his 1926 Beloved Belindy.

Erected during National Book Week in 1920, Marshall Field's "Volland Book House" was a popular stopping-off place for children and their parents, and an integral part of Volland's merchandising plan. (Marshall Field's)

Sometimes we half suspect Johnny is one of those little gnomes we read about—he seems to know so much about fairies of all kinds and where they live and what they do.

1920s Volland Company flyer

these books were several new Gruelle-designed toys based on protagonists in his *Little Brown Bear, Eddie Elephant,* and *Sunny Bunny* "Sunny Books."

With solid sales and steadily increasing annual royalties, Johnny Gruelle had every reason to feel successful. However, the charming little dolls that were putting food on his family's table and a sizeable nest egg in his bank account were also at the root of Gruelle's growing dissatisfaction.

Concerned about the quality of the Muskegon Toy-Company made Raggedy Ann dolls and desiring more direct control over manufacturing, Gruelle in 1921 had entertained an offer from Ross & Ross (the company that produced the Sunny-Book-based character dolls) to take over the manufacturing of both Raggedy Ann and Andy, supplying them directly to Gruelle rather than to Volland. However, Gruelle's restrictive contracts (in which he had granted the Volland Company exclusive marketing rights to his rag dolls for as long as it remained publisher of his books) pre-empted such a deal.

Furthermore, Volland's dwindling sales team and the fact that several years had passed since a new Raggedy Ann title had been published gave Gruelle

GUESS WHAT'S WRITTEN ON MY HEART? THE VOLLAND VALENTINES

The Volland Company was known not only for its elegant books; it also had earned a reputation for fine greeting cards, many designed by its capable illustrators. In August 1919, Johnny Gruelle supplied Volland with twenty-four different designs for valentines. Six of these cards, issued in 1920, were Raggedy Ann designs. The remaining eighteen were based on Gruelle's *Funny Little Book* and *Sunny Bunny,* as well as on Elizabeth Gordon's *Flower Children,* although no examples of these valentines could be documented. In late 1920, Gruelle created a new set of Raggedy valentines, which were issued in 1921. A selection is noted below:

"To My Valentine" Raggedy Ann (1920) (#4228)
"Good Morning! . . ." Raggedy Ann (1920) (#4229)
"Raggedy Ann and I hope . . ." Raggedy Ann (1920) (#4230)
"Just One Guess . . ." Raggedy Ann (1921) (#7426)
"Oh, Valentine!" Raggedy Ann (1921)
"A Valentine Message . . ." Raggedy Ann (1921) (#7429)
"I'm Looking for a Valentine" Raggedy Andy
"I Wonder . . ." Raggedy Ann (1921) (#7431)
"Be My Valentine . . ." Raggedy Andy (1921) (#7433)
"Guess What's Written on My Heart . . ." Raggedy Andy (1921) (#7434)
"I'm Looking for a Valentine . . ." Raggedy Andy (1921) (#7435)

(66)

(67)

(68)

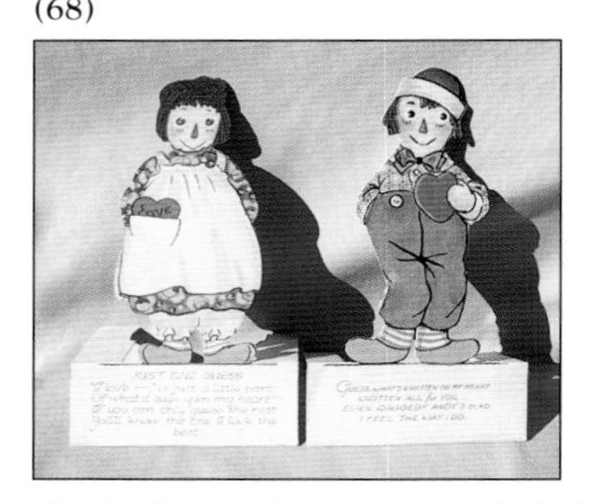

(69)

The Volland valentines (for which Gruelle was paid $10 per design) appeared both in die-cut and postcard form. On some, Raggedy Ann was given a tuft of real yarn hair. (Barbara Lauver/Brenda Milliren/Author's Collection)

even more reason to question the company's commitment to his dolls and books. Resentful of Volland's lukewarm attitude and iron-clad hold on his character properties, Gruelle concurred with his attorney, John Thompson, who had observed that Volland officials were "working very close to the edge on the fifty-fifty agreement so far as you desire to be paid a decent sum for your work is concerned. It is rather insulting to one's intelligence."

In spring of 1923, Johnny Gruelle took a break, embarking on a cross-country working vacation. Traveling with his wife, Myrtle, and two sons, Worth and Dickie, in a specially outfitted bus,

I met so many children out in Indianapolis who spoke of Johnny Mouse and the Woozgoozle the first thing and who liked the stories better than the Ann stories.

Johnny Gruelle to the Bobbs-Merrill Company
October 1921

Gruelle had as his ultimate destination Ashland, Oregon. Basking in the balmy climate and relaxed lifestyle of the West but intent on working, Gruelle continued generating his pithy "bird's-eye-view" cartoons for *Judge* magazine, as well as daily "Raggedy Ann and Raggedy Andy" newspaper serials. All the

(continued on page 75)

FOR GOOD LITTLE BOYS AND GIRLS: THE JOHNNY MOUSE DOLL

The sprightly little Johnny Mouse dolls were a planned tie-in to Gruelle's "Johnny Mouse and the Woozgoozle" stories that appeared in *Woman's World* magazine beginning in June 1920. Based on Gruelle's patented design and manufactured by Ross & Ross, these 10" felt-and-muslin stuffed dolls came costumed in red felt pants, print cotton shirt, and felt collar. The dolls were sold in *Woman's World* as $1 mail-order items during 1921.

In 1922, the Bobbs-Merrill Company considered introducing a new Johnny Mouse doll as a tie-in to Gruelle's forthcoming *Johnny Mouse and the Wishing Stick* book, but (as it had with an earlier idea for an Orphant Annie doll) the company scrapped the plan. *Woman's World* took advantage of the book's publication, however, again advertising Johnny Mouse dolls in its November and December 1922 issues, this time at a reduced price of 75 cents.

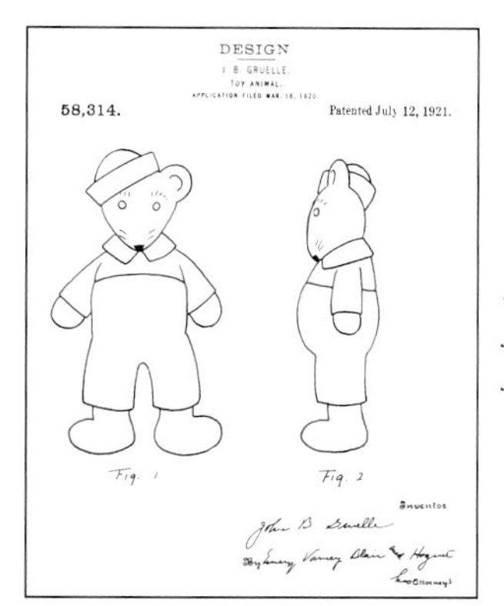

A fourteen-year patent for Johnny Mouse was registered July 12, 1920.

Johnny Gruelle's New Johnny Mouse Doll for Woman's World Girls and Boys

75c Postpaid

JOHNNY MOUSE stands ten inches high, wears red felt pantaloons, percale blouse, black gaiters, yellow tie, and a little yellow hat that fits over one ear and makes you laugh just to look at it. His little black eyes fairly sparkle with mischief and good nature. His very presence is a guarantee of happy days ahead. You can treat him rough, too, and he won't lose his temper or shape.

Sent prepaid while the supply lasts for only 75c—or free for 3 yearly subscriptions at 50c.

WOMAN'S WORLD
107 South Clinton St. Chicago

(70)

The felt-and-muslin Johnny Mouse Dolls stood 10" tall, had shoe-button eyes and a string tail, and wore red felt pants, percale blouse, black gaiters, and yellow tie and hat. The dolls were advertised in the fall 1921 and 1922 issues of Woman's World. *(Barbara Lauver/Author's Collection)*

(71)

THE GOLDEN PENNY

A Johnny Mouse and The Woozgoozle Story

JOHNNY GRUELLE

(72)

(Above) Gruelle's Woman's World *serial, "Johnny Mouse and the Woozgoozle," which ran June 1920-September 1921. (Right) Illustrated endpapers from* Johnny Mouse and the Wishing Stick, *published by Bobbs-Merrill (1922).*

(73)

As tie-ins to Gruelle's story series, Woman's World *offered an entire line of fabric goods stamped with images of Johnny Mouse and the Woozgoozle for home seamstresses to embroider and finish.*

MERRY LITTLE FRIENDS OF OURS: THE ROSS & ROSS DOLLS

After plans fell through in 1920 for Beers-Keeler-Bowman to produce toys based on Gruelle's patented Eddie Elephant and Henrietta Hennypenny characters, Johnny turned to the Ross & Ross Company of Oakland, California, to manufacture character dolls based on several of his Volland Sunny Book characters.

While the Volland Company would be primary retailer for these new dolls, selling them in tandem with the Sunny Books that had inspired them, Gruelle had negotiated the doll deal directly with Ross & Ross. This move not only gave him more control over design and quality; it also earned him a 10 percent royalty—double the 5 percent he would continue to receive from Volland for the Raggedy Ann dolls.

Manufacturing began in 1921, when Ross & Ross began producing a 12" plush Little Brown Bear doll, followed in 1922 by a 12" plush Sunny Bunny and 14" felt-and-cotton Eddie Elephant—all three based on Gruelle's homemade prototypes. Each fully costumed, kapok-stuffed toy came with a cloth name-and-trademark tag and was sold in its own box. The Ross & Ross dolls were manufactured until 1923. In all, 8,700 Little Brown Bears, 1,800 Eddie Elephants, and 5,400 Sunny Bunny dolls were produced.

Gruelle's design for Eddie Elephant echoed this newspaper cartoon rendered a decade earlier.

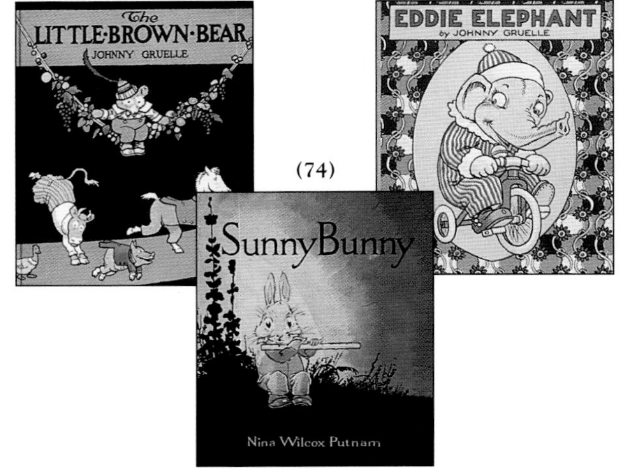

(74)

The books that inspired the dolls.

The U.S. Patent Office required Gruelle to change his original designation "Doll" to the more descriptive "Toy Animal" before registering his design patents for Eddie Elephant and Little Brown Bear.

(75)

The Ross & Ross Sunny Book dolls had sewn-on costumes with cloth tags asserting their patented, rather than generic, designs. Volland purchased the dolls from Ross & Ross for around $1 each. Top hat in this example is not original. (Jacki Payne/ Author's collection)

Gruelle's animated illustrations for J. P. McEvoy's The Bam Bam Clock *(1920) depicted dainty fairies, lush gardens, and vivid home interiors.*

Volland illustrator Marie Honre Meyers included Raggedy Ann and Andy in her revised edition of The Jolly Kid Book. *(The Chicago Public Library)*

Gruelle created this elf-and-fairy filled illustration for the January 1921 issue of John Martin's Book, *fulfilling the magazine's stated mission to present uplifting, enriching material for children.*

His little black eyes fairly sparkle with mischief and good nature. His very presence is a guarantee of happy days ahead. You can treat him rough, too, and he won't lose his temper or his shape.

Johnny Mouse doll ad
Woman's World
November 1922

(76)

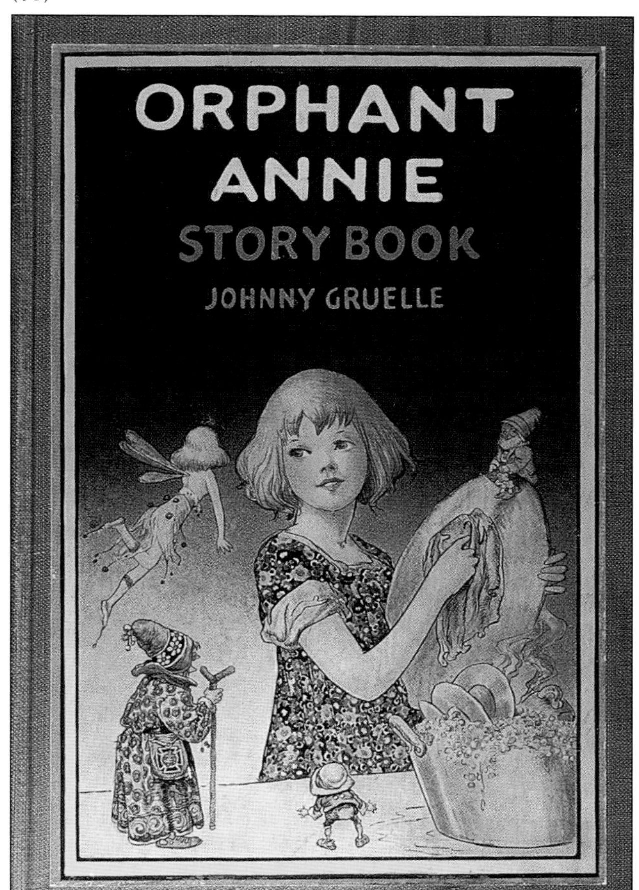

In 1921, Gruelle was commissioned by the Bobbs-Merrill Company of Indianapolis to write and illustrate Orphant Annie Story Book, *a tribute to the late Indiana poet James Whitcomb Riley. A year later the company would publish Gruelle's* Johnny Mouse and the Wishing Stick.

continued from page 70

while he was completing two new books for Volland—*Raggedy Ann and Andy and the Camel with the Wrinkled Knees* and *Raggedy Andy's Number Book*.

When Gruelle and his family finally returned to Norwalk in November of 1924, Volland president F. J. Clampitt had announced his retirement. The company had been absorbed by Gerlach-Barklow, an art-calendar-and-greeting-card enterprise based in Joliet, Illinois. Though the Volland name and product lines would be retained, the acquisition resulted in a headquarters move to the smaller Illinois city. Under its new ownership, Volland began issuing at least one Raggedy book a year and continued placing orders for its Raggedy dolls, keeping them in active production.

The Raggedy Ann and Andy dolls continued to be produced for Volland by different companies, and in 1928, Gerlach-Barklow even entered the fray as a doll manufacturer. Whether this move was an experiment in consolidation, a way to eliminate overhead, an effort to exercise more control, or because demand for dolls exceeded the production capacity of outside manufacturers is anybody's guess. In any event, years later a former art director at "the Gerlach" vividly recalled staff excitement when Raggedy doll fabrication began taking place on the top floor of the Joliet plant in a room full of whirring sewing machines.

"Oh!" the Camel with the Wrinkled Knees laughed, as Raggedy Ann tied her pocket hanky over the Camel's shoe button eyes, "Now I see!"

Raggedy Ann and Andy and the Camel with the Wrinkled Knees
1924

Though Gruelle remained skeptical of his publisher, for several years things seemed on track with Raggedy Ann and Andy, even after Wall Street's crash in 1929 ushered in the Great Depression. Many American publishers and doll and toy companies turned to manufacturing low-end merchandise to be sold in five-and-dime stores; many eventually went out of business altogether. However, for the time being, Volland remained committed to its high-quality books and merchandise. In spring of 1931, the company even launched an impressive line of fabric, plush, and felt character dolls, most of them inspired by Gruelle book characters.

continued on page 83

In addition to creating a dozen brilliant color plates, Johnny Gruelle crafted numerous fluid pen-and-ink drawings for The Magical Land of Noom *(1922).*

"RAGGEDY ANN" DOLLS

By KATHLEEN ASP

ONE of my third-grade pupils brought the book "Raggedy Ann" to school and asked me to read it to the children. After the book was read the boys asked if they might make a "Raggedy Ann" doll. I cut a pattern to represent "Raggedy Ann" from the picture on the front of the book and from this the children made their dolls.

The doll when completed is about twelve inches long. The body is made first; then the legs are stuffed and sewed on to the body; the shoe button eyes are added; the yarn hair is sewed on; and the face painted. Each child cut his doll's dress and apron and sewed them by hand.

Our rule was that all school work must be finished before the pupils could sew on their dolls. The school work showed much advancement during the two weeks we made "Raggedy Ann" dolls. Later we made "Raggedy Andy," Ann's brother, in the form of a jumping jack. For composition work, each child wrote a story of "Raggedy Ann."

Normal Instructor, *January 1925. (Candy Brainard)*

ADVENTURES OF RAGGEDY ANN AND RAGGEDY ANDY

By Johnny Gruelle

"LOOK at that funny little woman!" Raggedy Andy whispered to Raggedy Ann.

"I've been watching her come along the path," Raggedy Ann replied. "Something seems to be bothering her."

The queer little woman was only about an inch and a half high and came walking along in a jerky sort of way until she came to Raggedy Ann's foot. Then looking up she saw the two dolls smiling at her.

"Why are you laughing at me

when my finger hurts so much?" the little woman asked.

"We can't help it!" Raggedy Ann replied. "Our smiles are painted on and can't come off until they wear off!"

"Well! My finger hurts very, very much!" the queer little woman said.

"We are very, very sorry!" Raggedy Ann said. "How did you hurt your finger?"

"I live over in the deep Woods

and it makes me very angry!"

"Do they make the rain wate muddy?" asked Raggedy Ann.

"Of course not!" the queer littl woman answered. "But it is my clam shell and I don't want anyon else to use it!"

"Hmm!" Raggedy Ann mused "and you hurt your finger on the clam shell?"

"No, I didn't," the queer littl woman answered. "I caught one o the Elfin boys swimming in my pon and I shut him up in an acorn fo a week. Then Helen Honeybee cam and let him out, and when I struc her, I ran a needle in my finger; an it's been in there for a week!"

"Well, well!" Raggedy Ann laughe softly down in her cotton stuffe throat. "Take the rag off your finge Mrs. Snoopy, and maybe we can hel you!"

Snoopy Doodjinipper unwrappe her finger and climbed upon Raggedy Ann's apron, so that the tw rag dolls could see. "Yes, the needl is still in there," said Raggedy Ann "but we cannot pull it out becaus we have no fingers on our hands!"

"Oh, dear, what shall I do?" waile Snoopy. "It feels like it is goin 'bump, bump, bump!' all the time!"

Just then Percy Pinchingbu climbed upon Raggedy Ann's apron and Ann clapped her hands togethe so suddenly, Snoopy Doodjinipper was startled and fell over backwards. "Here, Percy Pinchingbug!" said Raggedy Ann, "will you please pull this needle from Snoopy's finger?"

Percy Pinchingbug got out his pinchers and soon pulled Helen Honeybee's needle from Snoopy's finger.

"Oh, now it doesn't hurt at all!" said Snoopy Doodjinipper. "Thank you so much, Raggedy Ann and Percy Pinchingbug!"

"You're quite welcome," said Raggedy Ann.

"You are welcome," said Percy Pinchingbug. "Little weeny, teeny Eddy Tif sent me to pull it from your

"I wrote one thousand daily stories, <entitled> 'Raggedy Ann,' for United Features," recalled Gruelle about these newspaper serials that earned him as much as $50 per episode and served as source material for most of his later books.

(77)

Volland Raggedy Andys: (left to right) "Single Eyelash," "Crescent Smile," "Crescent Smile," "Transitional," "Transitional," and "Finale." (Candy Brainard)

(78)

Volland Raggedy Anns: (left to right) "Patent," "Patent," "Single Eyelash," and "Finale." (Candy Brainard)

STEPPING STONES: THE RAGGEDYS ONSTAGE

When stars Fred and Dorothy Stone loped onstage as Raggedy Ann and Andy in the 1923 Broadway musical "Stepping Stones," Gruelle's little rag dolls were an instant hit. By 1924, "Stepping Stones" was playing to capacity crowds across the U.S., and Raggedy Ann and Andy had endeared themselves to an avid, theater-going audience. The production's catchy songs, written by Anne Caldwell and Jerome Kern, were published as sheet music and even inspired a fashionable foxtrot-style dance known as "the Raggedy Ann." All this activity gave an added boost to already- brisk sales of Gruelle's books and dolls.

Like other children's authors before him (including L. Frank Baum of *Oz* fame), Gruelle relished seeing his book characters brought to life on the stage. However, "Stepping Stones" was neither the first nor the last time the Raggedys were invited to star in a theater production.

Offices of
Charles Dillingham
Globe Theatre
Broadway & 46th Street
New York

October 9, 1923.

Mr. Johnny Gruelle,
114 Granite Street,
Ashland, Oregon.

Dear Mr. Gruelle:

As the number develops at rehearsals we are all getting to be RAGGEDY ANN fans. I have a new idea to propose to you. What would you think of letting us have the rights to make an entire musical comedy with RAGGEDY ANN and RAGGEDY ANDY as the principal characters? If this strikes you favorably will you give us an option on the stage rights so that I may have something to work with in getting Jere Kern interested in writing the music? You said in your last letter that you were settling down for a while, but if you were coming East in the near future for any other reason we could talk it over so much better than through correspondence.

With best wishes.

Yours sincerely,

During rehearsals for "Stepping Stones," Charles Dillingham's office wrote this note to Johnny Gruelle. (Cox-Scheffey Collection)

Fred and Dorothy Stone portrayed Gruelle's floppy rag dolls in the Broadway musical "Stepping Stones."

(79) (80)

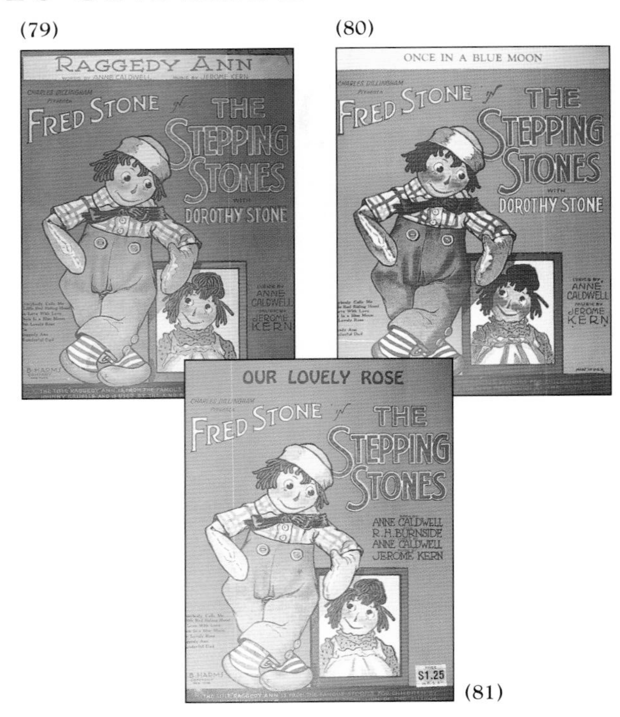

(81)

In 1923 and 1924, the T. B. Harms Company of New York published sheet music from "Stepping Stones." "Raggedy Ann" was recorded multiple times and even issued as a piano roll by the Mills Novelty Company.

Less than a year after Raggedy Ann had made her 1918 book debut, Joseph M. Gates had approached Gruelle, asking him to consider a Raggedy Ann musical. "Stepping Stones" grew from this seed, and following its success, Gruelle would continue to receive requests to use his Raggedy characters in everything from elaborate stage productions to modest puppet plays. He usually granted permission.

Where the Volland Company was concerned, however, Gruelle was more protective. Having agreed to share dramatic and moving picture rights with Volland for anything based on his *Raggedy Ann Stories*, Gruelle wisely retained exclusively for himself the rights to all subsequent Raggedy Ann and Andy books. Gruelle remained hopeful that he would someday collaborate with a famous composer to create the quintessential Raggedy Ann stage production.

Images of Raggedy-style dolls helped sell 1920s foodstuffs: (above) Poster *maga-zine, September 1925; (at right)* Holland Magazine, *March 1925.*

(82)

I love a lady who has black shoe button eyes,
I love a lady with a rag bag in her head;
In a parade of dumbbells she would take the prize;
But when it comes to ragtime she would knock 'em dead.

"Raggedy Ann"
Anne Caldwell and Jerome Kern
1924

In 1929, discounted Raggedy Ann and Andy dolls were offered by the Sears, Roebuck catalogue. Despite changes in manufacturers and fluctuations of retail prices, Gruelle continued to receive (as he had since 1918) a modest 5-cent royalty for each Raggedy Ann or Raggedy Andy doll sold.

(83)

The Volland Raggedy Ann and Andy bookshelf, advertised in The Paper Dragon *(1926), measured approximately 18"(h) x 20"(w) and featured Raggedy Ann and Andy peeking over the top and at either end. (Patricia Caze Deig)*

TO FILL A CHILD'S ARMS: THE VOLLAND BELOVED BELINDY DOLL

On July 1, 1926, Johnny Gruelle authorized the Volland Company to begin selling a stylized African-American doll as a tie-in to his forthcoming book, *Beloved Belindy*. In creating a Beloved Belindy character, Gruelle and his publisher were hoping to tap into the growing popularity of a folk-styled "mammy" doll that had originated in the southern United States.

Sturdy and voluptuous, the hand-wrought Volland Beloved Belindy stands 15" tall. Constructed of dark brown cloth with red-and-white legs and red fabric feet, Beloved Belindy has hand-painted lips, eyebrows, and nose (though some later examples appear to be machine-stamped) and white shirt-button eyes. Her clothing consists of a floral patterned skirt, a solid red blouse, white apron and pantalettes, and a red bandanna in lieu of hair. Although no sales or manufacturing records could be documented, the Volland Beloved Belindy doll was advertised and sold between 1926 and 1934.

Interestingly, Gruelle did not register a design patent for his Beloved Belindy doll. He and Volland may have erroneously assumed that the book copyright for *Beloved Belindy* would also cover a plaything of the same name and design. Ironically, even the book copyright would not be registered until 1941.

Though by today's standards unacceptably stereotypical, the Volland Beloved Belindy doll was designed by Gruelle to appear both commanding and nurturing—completely in keeping with the amiable but assertive persona he had created for her in his book.

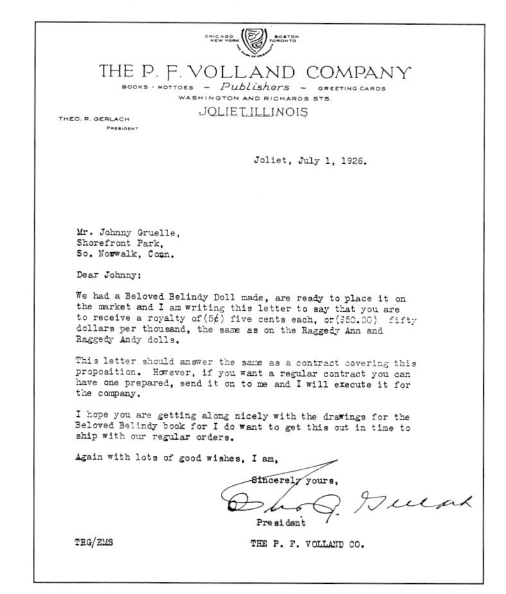

CHICAGO NEW YORK BOSTON TORONTO

THE P. F. VOLLAND COMPANY

BOOKS · MOTTOES – *Publishers* – GREETING CARDS

WASHINGTON AND RICHARDS STS.

JOLIET, ILLINOIS

THEO. R. GERLACH
PRESIDENT

Joliet, July 1, 1926.

Mr. Johnny Gruelle,
Shorefront Park,
So. Norwalk, Conn.

Dear Johnny:

We had a Beloved Belindy Doll made, are ready to place it on the market and I am writing this letter to say that you are to receive a royalty of (5¢) five cents each, or ($50.00) fifty dollars per thousand, the same as on the Raggedy Ann and Raggedy Andy dolls.

This letter should answer the same as a contract covering this proposition. However, if you want a regular contract you can have one prepared, send it on to me and I will execute it for the company.

I hope you are getting along nicely with the drawings for the Beloved Belindy book for I do want to get this out in time to ship with our regular orders.

Again with lots of good wishes, I am,

Sincerely yours,

President

TRG/EMS THE P. F. VOLLAND CO.

Ted Gerlach's letter to Gruelle was intended to suffice as a contract for the Volland Beloved Belindy doll. (Cox-Scheffey Collection)

(84)

(85)

(86)

Gruelle cast his Beloved Belindy character as a wise and kindly leader of the playroom—the role usually reserved for Raggedy Ann. Volland paid Gruelle a 5-cent royalty for each Beloved Belindy doll sold. (Candy Brainard/Author's Collection)

A Volland Beloved Belindy and her young friend in the late 1920s. (Cynthia Dorfman and Michael Patrick Hearn)

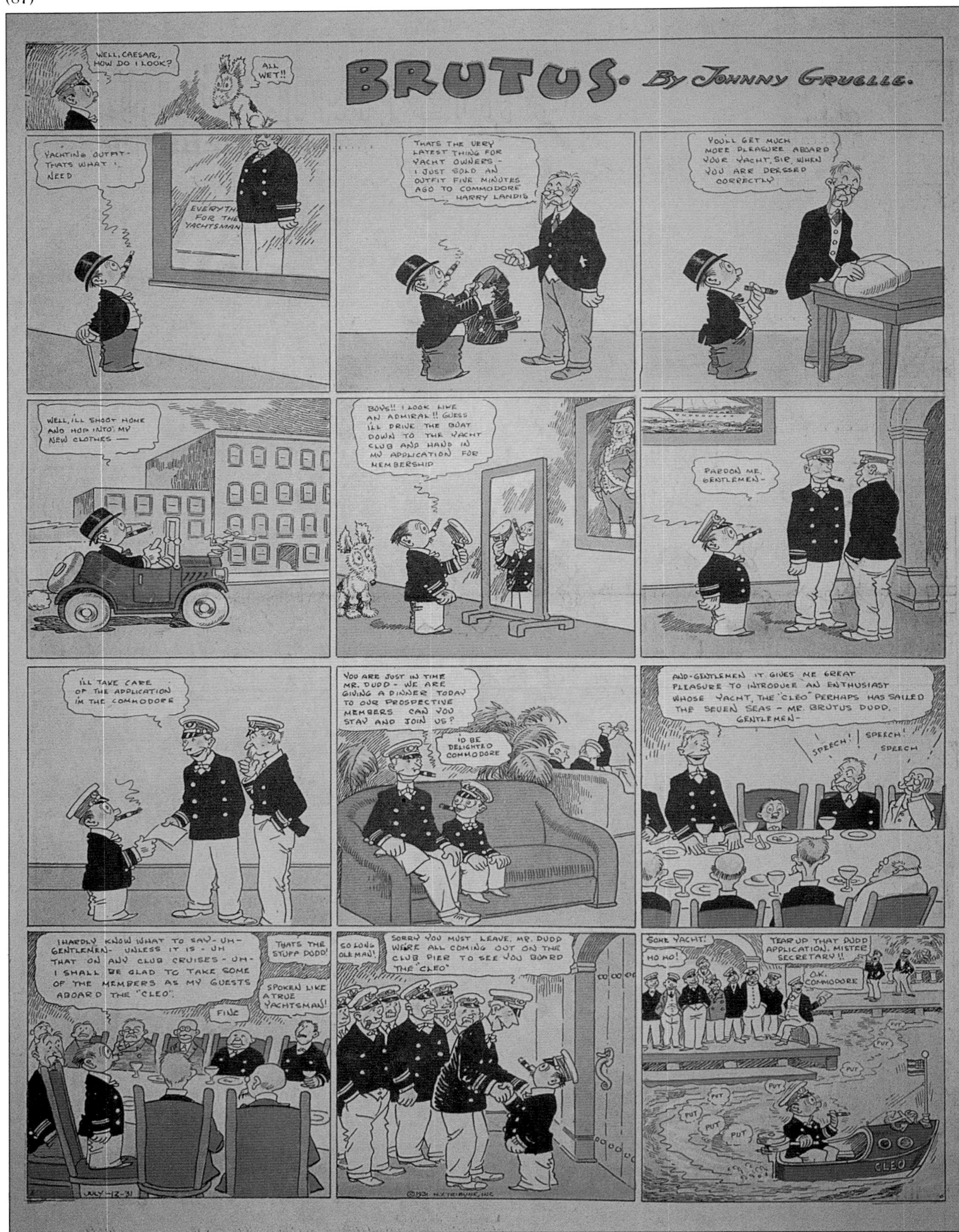

Gruelle's "Brutus," appearing in the New York Herald Tribune *between 1929 and 1938, was the cartoonist's semi-autobiographical take on a Depression-era man's world.*

During the late 1920s, Pictorial Review *offered Raggedy costume patterns for home seamstresses. (Marge Meisinger)*

continued from page 75

In the meantime, spurred on by the mood of the Depression, Gruelle sought out other creative sources of income. He launched a new syndicated comic page for adults called "Brutus." Though not a trained musician (but a dabbler at piano and saxophone), he offered himself as a song lyricist, collaborating with composers such as Johnny Mercer, M. K. Jerome, Harry Archer, and Joan Jasmyn on 1930s-style "moon-croon-June" love ballads. Several, including "Lonely for You" and "Beyond the Moon," were published by T. B. Harms, publisher of "Stepping Stones."

In 1930, former Harms staffer Charles Miller teamed Gruelle with Will Woodin (a part-time composer and industrialist who later would serve as U.S. secretary of the treasury under Franklin Roosevelt). The colorful result was *Raggedy Ann's Sunny Songs,* which Miller Music published (in 1937, a second Miller folio, *Raggedy Ann's Joyful Songs,* would appear, co-written by Gruelle and Miller, with illustrations by Gruelle).

Truly dearest, skies are clearest
When I'm with you.
My heart's singing, ting-a-linging
With love so true.

"Lonely for You"
Johnny Gruelle and Harry Archer
1930

In 1932, Gruelle moved with his family to Miami Beach, settling into a gracious, well-appointed home on the Nautilus Waterway and a lifestyle well out of reach for most during the Great Depression. By summer of 1934, following the death of its president, T. R. Gerlach, the Gerlach-Barklow Company finally succumbed to hard times and declared its Volland subsidiary out of the doll and book business altogether. Assigning rights back to Gruelle, Gerlach-Barklow began remaindering its Raggedy book and doll inventory, which had been discounted to well below wholesale, yielding no further royalties.

Not only had Gruelle lost a publisher and doll purveyor, he would now be without the steady and reliable income he had depended on, an average of $20,000 annually. Furthermore, it was now solely up to him to protect his trademarks and keep his dolls and published works available for a buying public.

continued on page 88

(88)

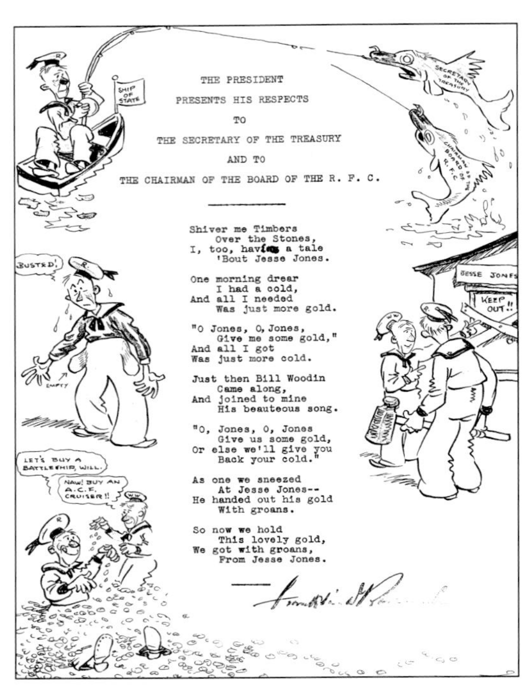

THE PRESIDENT
PRESENTS HIS RESPECTS
TO
THE SECRETARY OF THE TREASURY
AND TO
THE CHAIRMAN OF THE BOARD OF THE R. F. C.

Shiver me Timbers
Over the Stones,
I, too, having a tale
'Bout Jesse Jones.

One morning drear
I had a cold,
And all I needed
Was just more gold.

"O Jones, O, Jones,
Give me some gold,"
And all I got
Was just more cold.

Just then Bill Woodin
Came along,
And joined to mine
His beauteous song.

"O, Jones, O, Jones
Give us some gold,
Or else we'll give you
Back your cold."

As one we sneezed
At Jesse Jones--
He handed out his gold
With groans.

So now we hold
This lovely gold,
We got with groans,
From Jesse Jones.

In this June 1930 "Niles Junction" cartoon for College Humor *magazine, Gruelle paid tribute to his longtime friend and business associate Howard Cox.*

Johnny Gruelle collaborated with Pres. Franklin D. Roosevelt and Sec. of the Treasury Will Woodin on poems spoofing Texas mogul Jesse H. Jones. (Franklin D. Roosevelt Library)

(89)

The Volland Company issued boxed sets of lithographed puzzles featuring Gruelle endpaper artwork from its "Happy Children" and "Sunny Book" series. Among them were "Sunny Days for Sunny Children" and "Johnny Gruelle's Happy Children" (shown here). (Candy Brainard)

With his love for the performing arts, it was a natural that Gruelle would collaborate on musical compositions like these from the 1930s.

WE'RE HERE TO GREET YOU! THE VOLLAND CHARACTER DOLLS

Following several months of targeted advertising, in May 1931 the Volland Company introduced ten Gruelle-inspired character dolls as part of a fourteen-item line collectively referred to as the "Raggedy Dolls." Included were Raggedy Ann and Andy and Beloved Belindy; a newly designed Sunny Bunny, Eddie Elephant, and Little Brown Bear; and a first-time-ever Uncle Clem, Percy (the) Policeman, Eddie Elf, and Pirate Chieftain.

The characters appeared in Gruelle's *My Very Own Fairy Stories*, *Wooden Willie*, *Raggedy Ann in the Deep Deep Woods*, *Raggedy Ann and Andy and the Camel with the Wrinkled Knees*, and *Marcella*. Rounding out the line were four non-Gruelle dolls: Brownie Tinkle Bell and Bunnie Bull (introduced as "two surprise dolls"), and Pinky Pup and Empty Elephant (characters from *Pinky Pup* and *Pinky Pup and the Empty Elephant*, Volland books by Dixie Willson and Erick Berry).

(91)

(92)

The Volland Uncle Clem doll (bottom left) had shoe button eyes and came dressed in a gingham "kilt" and felt hat. Gruelle may have gotten the idea for his Uncle Clem doll (which first appeared in his 1917 book My Very Own Fairy Stories*) (top) from a 1913 Steiff Scots doll (bottom right) called "Scott." (Brenda Milliren/Strong Museum/Author's Collection)*

(93)

(94)

(95)

The Volland Percy (the) Policeman character doll (top) was first depicted in My Very Own Fairy Stories *(bottom). This doll may have been inspired by Steiff's "American Policeman" doll (middle left) or by one of many turn-of-the-century generic felt dolls (middle right). (Author's Collection/Pat Planton/Strong Museum)*

The Volland character dolls have no company markings or body tags, but each was issued with a paper hang tag identifying it by name. Though their manufacturer remains unknown, the dolls could have been produced by the C. B. Moore Company, which by that time was Volland's primary doll manufacturer.

The Volland character dolls retailed for $2.00 (Sunny Bunny and Little Brown Bear, for $2.50); by 1932, prices were reduced by 50 cents. Gruelle received a 5-cent royalty for each doll.

(96)

Special painted benches, with the logo "Reserved for Raggedy Dolls," were offered to retailers who ordered full sets of the Volland character dolls.

(97)

The Volland Pirate Chieftain (portrayed by Gruelle in Raggedy Ann and Andy and the Camel with the Wrinkled Knees*) stood approximately 18" tall, had a shaggy yarn beard, and was dressed in felt clothing. (Candy Brainard/Author's Collection)*

Based on a character appearing in Wooden Willie *(1927) and* Raggedy Ann in the Deep Deep Woods *(1930), the 18" Volland Eddie Elf doll was also reminiscent of Gruelle's earlier Mr. Twee Deedle character. (Barbara Lauver/Author's Collection)*

(98)

Each Volland character doll came with a paper hang tag. (Barbara Lauver)

(99)

The Volland Sunny Bunny doll had auburn plush head and paws and a horizontal (rather than V-shaped) stitched nose. He wore a red coat, yellow vest, green trousers, and black hat. (Barbara Lauver)

(100)

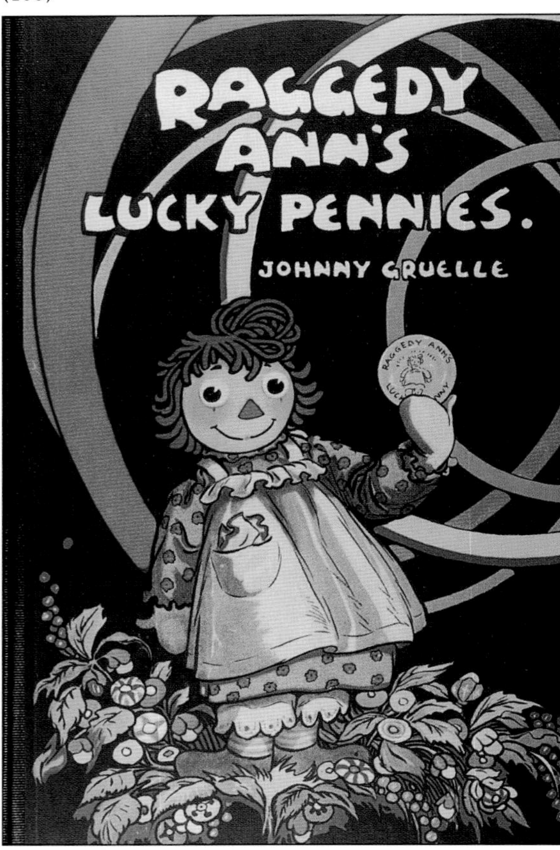

Johnny Gruelle, vacationing at Lake Sissabagami, Wisconsin, in the early 1930s. While there, Gruelle completed Raggedy Ann's Lucky Pennies *(1932), his final book for the Volland Company. (Betty MacKeever Dow/Author's Collection)*

continued from page 83

At least a part of the problem was solved when M. A. Donohue & Company of Chicago straightaway purchased reprint rights to most of Gruelle's Volland titles. Determined to get off on the right foot, company president Michael Donohue assured Gruelle, "We are looking forward to a relationship with you and feel that we can work out a program which will be mutually pleasant and profitable."

Meanwhile, Gruelle had been mulling over some more expansive options. During his years spent affiliated with the Volland Company he had learned the importance of protecting the earning potential of his characters. So when letters had begun arriving in 1932 from an old friend and colleague, Howard Cox, suggesting that Johnny form a business of his own to directly oversee all of his publishing, doll, and merchandising projects, Gruelle was intrigued.

Yoo, hoo! We're having a grand time these nights—Raggedy Andy and Beloved Belindy and I. 'Cause why? 'Cause we have a whole crowd of new dolls to play with. And what do you think? They are right out of the books about me. And the way Sunny Bunny and Eddie Elephant and Little Brown Bear romp about, you'd think they never had any fun before. But they have. There's a whole Sunny Book about each one of them. Bushels of Love, Raggedy Ann.

Volland Company promotional letter 1931

THE MADMAR PUZZLES

By the early 1930s, the Volland Company had turned over production of its boxed jigsaw puzzles to Madmar Quality Company of Utica, New York. Known for its accurately printed, up-to-date map puzzles and its "Picture Puzzle" series, Madmar was well respected as a wholesaler/retailer of high-quality educational merchandise.

Like the Volland puzzles, the Madmar puzzles were reprints of endpapers from Gruelle's Volland books. Issued in the company's Group "A" series (for children ages 3-7) and Group "B" series (for children ages 7-12), the Madmar puzzles are as follows:

Raggedy Ann Series (#505)
Six pictures assorted (including endpaper artwork from *Wooden Willie*, *The Paper Dragon*, and *Raggedy Ann's Wishing Pebble*). Each reversible puzzle makes two pictures. Box measures 9¼" x 12¼". $1

(101)

A double-sided puzzle in Madmar's "Raggedy Ann" series.

Aunt Belinda <sic> Series (#304)
Six pictures assorted (including endpaper illustrations from *Beloved Belindy* and *Wooden Willie*). Two puzzles in each box. Boxes entitled "Aunt Belinda Picture Puzzle" and "Raggedy Ann Picture Puzzle." Box measures 6¼" x 8¼". 25 cents

Funny Story Series (#404)
Six pictures assorted (including endpaper illustrations from *Raggedy Ann's Alphabet Book* and *Sunny Bunny*). Each reversible puzzle makes two pictures. Box measures 6⅛" x 8⅛". Also issued as "Funny Land Picture Puzzle"(#321). 25 cents

Mystery Series (#490)
Three pictures assorted (including endpaper illustration from *The Cheery Scarecrow*). Each reversible puzzle makes two pictures. Box measures 7¾" x 12". 75 cents

(102)

(103)

(104)

Madmar's 1931 line of Gruelle puzzles. (Alice Fleming)

(105)

The Miller Music folios showcased Johnny Gruelle's lyrical and artistic talents while cashing in on Raggedy Ann's and Raggedy Andy's popularity.

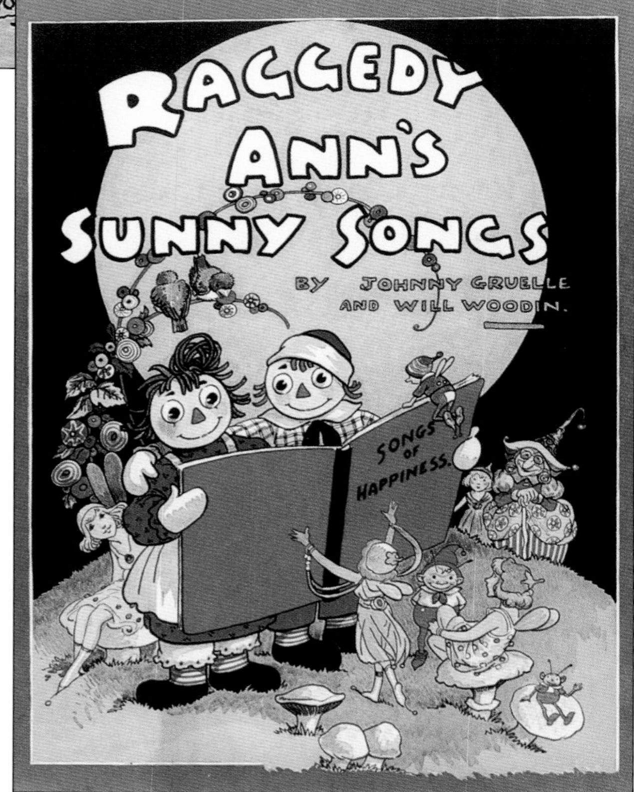

(106)

Cox had gone to work for Volland during the teens as a $150-per-week salesman and had been there when the company introduced Quacky Doodles, Danny Daddles, and Raggedy Ann and Andy. Departing in 1924 when Volland was absorbed by Gerlach-Barklow, Cox returned in 1927 to manage Volland's book division, eventually staffing the New York City office. The bond between Gruelle and Cox had grown over the years, nurtured by their Volland connections, similar senses of humor, and complementary temperaments—Gruelle, the retiring, free-spirited artist; Cox, the outgoing, enterprising businessman.

A consolidated business seemed the way to go, though the two had initially agreed not to do anything right away to disrupt Gruelle's income or upset his sensitive relationship with his publisher. However,

(107)

Cast-iron Raggedy Ann and Andy bookends (5½") were sold by the Volland Company in the early 1930s. Records indicate that a Mrs. Rosa May Pickard held the license for metal bookends. (Barbara Lauver)

THE DONOHUE BOOKS

Following the demise of the P. F. Volland Company in 1934, the Chicago-based M. A. Donohue & Company began reprinting most of Gruelle's Volland titles. By the 1940s, Donohue had put twenty Gruelle titles back in print—thirteen "Happy Children" books, six "Sunny Books," and *The Magical Land of Noom*. The company would continue its reprintings for more than two decades, though it would never publish any new books by Gruelle.

Fortunately, Donohue had acquired both original artwork and mechanical plates from Volland, resulting in "first-generation" reprints that were similar in quality and clarity to the original editions. Except for the Donohue imprint on the title page, the earliest Donohue reprints looked very much like the final Volland editions. However, as Donohue began standardizing certain features, its reprints took on their own distinctive look.

The first Donohue editions were bound in paper-covered boards, most with visible cloth spines (although a few appeared with paper-covered spines). Pages were printed on coated paper (matte or shiny), the Donohue imprint appearing on the title page along with the Volland triangular publisher's device. No editions or printings were stated. Instead of utilizing gift boxes, all Donohue reprints were issued in illustrated paper dust jackets.

A distinguishing feature of the Donohue reprints was their standardized endpapers. For "Happy Children" titles, the Volland "dolls and castle" endpaper from *Raggedy Ann's Magical Wishes* was used; for all six "Sunny Books," the Volland endpaper from *Little Sunny Stories* was used. (Note: Several endpaper anomalies have been noted in some very early Donohue reprints.)

Most Donohue editions include a distinctive last-page advertisement that lists currently available Raggedy Ann titles. In the "Happy Children" titles, the ad design was based on the back-page ad in Volland's *Raggedy Ann's Lucky Pennies*. In the "Sunny Book" reprints, the ad from each original Volland edition was used. The only change was the publisher imprint.

Donohue continued reprinting Gruelle's "Happy Children" books until the early 1940s when the

Though covers of most Donohue reprints resembled the original Volland editions, for one edition the company gave Raggedy Ann Stories *a revised title and new cover.*

(108)

(109)

During the 1940s, Donohue began issuing reprints of Gruelle's "Sunny Books" in solid red and orange covers with simple black line art.

(110)

Donohue sold its "Sunny Book" reprints individually and as a boxed set.

Johnny Gruelle Company began issuing its own reprints. However, four titles (*Wooden Willie*, *Raggedy Ann's Magical Wishes*, *Marcella*, and *Raggedy Ann's Lucky Pennies*) that were never reprinted by the Gruelle Company would continue to be issued by Donohue, which remained a Gruelle Company licensee until the 1960s.

By the late 1930s, Donohue "Sunny Books" were appearing with red or blue cloth-covered covers stamped in black only, which later gave way to orange paper-covered boards. In 1941, Donohue reprinted Gruelle's "Sunny Books" (all but *The Funny Little Book*) in the oversized *Johnny Gruelle's Golden Book*. Though Donohue was allowed to continue issuing *Johnny Gruelle's Golden Book* until at least 1953, by 1947 the company had surrendered individual reprint rights for all six "Sunny Books" to the Johnny Gruelle Company (which never went ahead with "Sunny Book" reprints of its own).

Around 1935, Donohue began reprinting Gruelle's *The Magical Land of Noom*. Though issued in dust jackets rather than gift boxes, the first Donohue *Noom* reprints resemble the original Volland editions in most other respects, including the illustrated cover label, distinctive map endpapers, and color plates. By the late 1930s, Donohue was issuing reprints of *Noom* in red cloth-covered boards stamped in black, an edition that remained available for well over a decade.

(111)

(112)

Early Donohue reprints of The Magical Land of Noom *had blue covers with illustrated cover labels; covers of later reprints were red with black line art.*

(113)

Donohue's 1944 catalogue featured the Raggedys on its cover.

The "Doll and Castle" endpaper used by Donohue in most of its "Happy Children" reprints came from the Volland edition of Raggedy Ann's Magical Wishes.

(114)

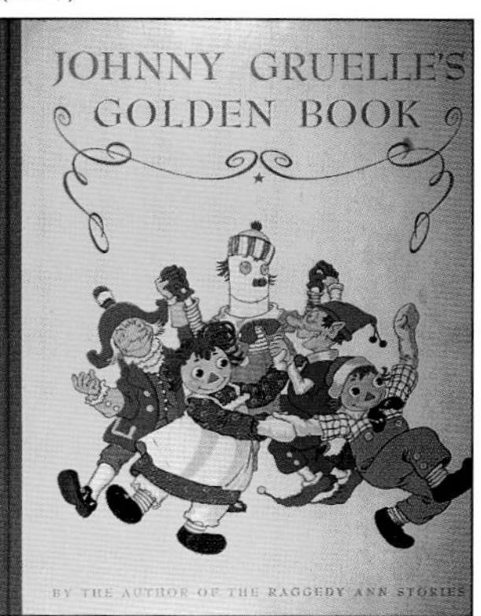

This oversized 1941 reprint of Gruelle's Volland "Sunny Books" (licensed to Donohue by Gruelle in the 1930s) ignited a controversy with the newly formed Johnny Gruelle Company over control of the "Sunny Books."

The December 1930 release of R.C.A. Victor's Raggedy Ann's Sunny Songs *was advertised in newspapers and specialty music publications.*

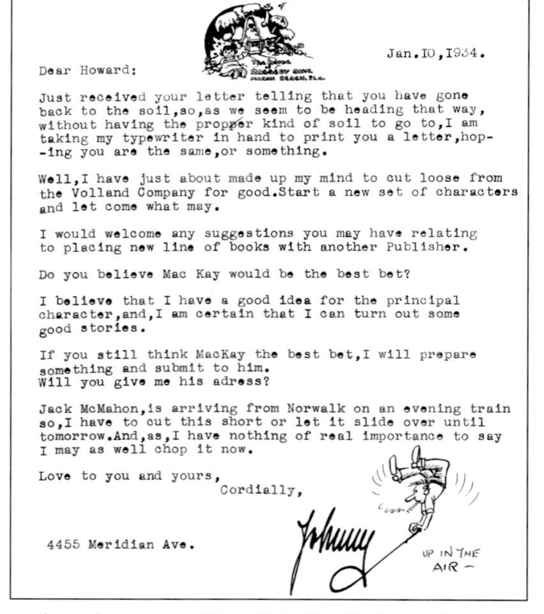

Jan.IO,I934.

Dear Howard:

Just received your letter telling that you have gone back to the soil,so,as we seem to be heading that way, without having the propper kind of soil to go to,I am taking my typewriter in hand to print you a letter,hop--ing you are the same,or something.

Well,I have just about made up my mind to cut loose from the Volland Company for good.Start a new set of characters and let come what may.

I would welcome any suggestions you may have relating to placing new line of books with another Publisher.

Do you believe Mac Kay would be the best bet?

I believe that I have a good idea for the principal character,and,I am certain that I can turn out some good stories.

If you still think MacKay the best bet,I will prepare something and submit to him.
Will you give me his adress?

Jack McMahon,is arriving from Norwalk on an evening train so,I have to cut this short or let it slide over until tomorrow.And,as,I have nothing of real importance to say I may as well chop it now.

Love to you and yours,
Cordially,

Johnny

4455 Meridian Ave.

UP IN THE AIR —

Inspired by Howard Cox's idea for a family-owned business, Gruelle told him so in words and pictures. (Cox-Scheffey Collection)

(115)

During the late 1920s and early 1930s, Gruelle drew "Yahoo Center" for the humor magazine Life. *This one, from 1933, featured the names of Volland Company president Ted Gerlach, composer Will Woodin, and Burdines, a stationer in Miami Beach where Gruelle made frequent presentations for children.*

after Cox left the crumbling Volland Company for good in early 1934, Johnny wrote to him seeking help in designing a publishing and merchandising operation that would allow Gruelle (with Cox's marketing skill and printing and production connections) to oversee, control, and enjoy the spoils of his many copyrighted Raggedy Ann and Andy works. Cox agreed, reasoning that such a venture could give Gruelle independence and would assure that "everything pertaining to Raggedy Ann will have a focal point."

Gruelle was delighted with Cox's interest, and in the months that followed, Johnny and his wife, Myrtle (whose keen business sense had made her a trusted advisor), mused along with Cox about "Raggedy Ann, Inc." In August 1934, Cox proposed

"You see, Sarah," Percy, the policeman doll said, "Beloved Belindy has thought of everything!"

Beloved Belindy
1926

a three-part start-up plan for the prospective company. The first phase would be devoted to a brand-new Raggedy book. During phase two, Cox would arrange for an East Coast doll manufacturer to begin production of new Raggedy Ann, Raggedy Andy, and Beloved Belindy dolls. In the third phase, Johnny would generate greeting card designs. Like many forward-thinking businessmen during the Depression, Cox was convinced that there would be a turn-around in financial conditions, and he wanted to be ready for the upturn. He suggested beginning right away but proceeding cautiously.

(116)

Gruelle created this souvenir program during the early 1930s for a yachting event hosted by the Committee of One Hundred.

Children!

Meet Johnny Gruelle

Creator of "Raggedy Ann" and "Raggedy Andy"

Tuesday, in the Book Shop

The hundreds of little boys and girls who have grown up with their "Raggedy Ann" and "Raggedy Andy" story books will want to come to this store Friday afternoon at 3.30 to meet Johnny Gruelle, who wrote all the stories and drew the pictures! And he will put his name in your "Raggedy Ann" and "Raggedy Andy" books. The stories of these two rag dolls are in the Book Shop at $1.25 each.

H. K. & Co., Books, First Floor.

Gruelle's appearances and booksignings were well-advertised in local newspapers. (Cox-Scheffey Collection)

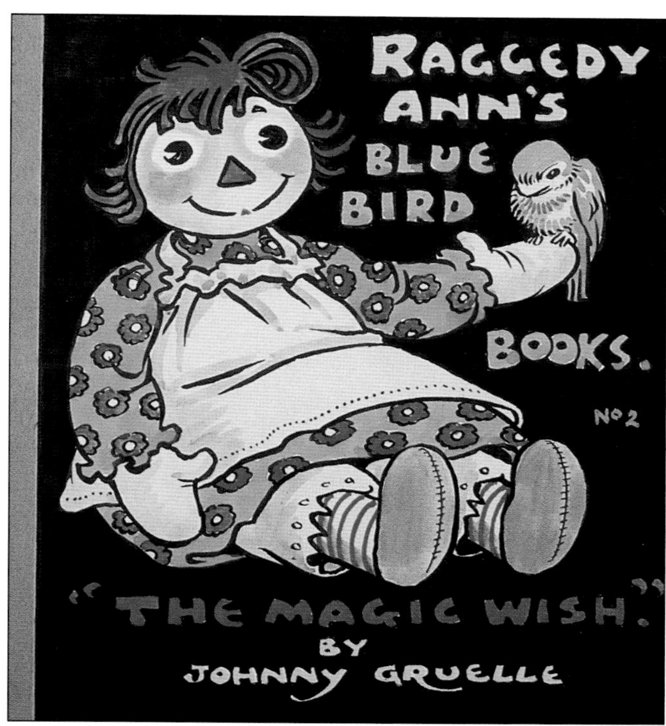

The Raggedy Ann "Bluebird Books," conceived by Gruelle in the 1930s to capture a mass-market audience, were never published. (Cox-Scheffey Collection)

Since the three dummy <"Raggedy Ann's Bluebird"> books reached me last Friday morning I have been digging up dope. It looks most interesting and I am pretty well steamed up over the idea that this thing can be put over in a really profitable way.

Howard Cox to Johnny Gruelle
June 1933

"There is no question that the Volland foolishness of the past three years has perceptibly hurt and it would be an easy thing to make the wrong movie or radio contract that would kill the whole thing," Cox observed. "I would feel justified in devoting the greater part of my time to manufacturing the products and selling them as well as handling the radio, picture, and dramatic developments."

Gruelle responded: "You know what I think of you, and that nothing would please me better than to work with you on a proposition where it would be possible for me to control the whole situation, and whereby we could both show the Volland Co. where they made a mistake in selling out."

Cox's offer to invest in the company underscored his commitment, but by the time the Gruelles lined

During the late 1920s and early 1930s, Gruelle presented Raggedy stories at schools and bookshops, all while drawing pictures at an easel with a fat black crayon. (AP Wirephoto)

(117)

(119)

(118)

Between 1934 and 1938, Gruelle generated more than a thousand illustrated proverbs, which he referred to as "doggerel philosophical Anns." He created as many as fifty at one sitting for the George Matthew Adams Syndicate, which placed them in daily newspapers across the country.

up their own $10,000 to invest they seemed reluctant and uncertain. Cox moved on to other ventures, and for the time being Raggedy Ann, Inc., was tabled.

The thought of launching his own business may have been too daunting for Gruelle. Or perhaps he had decided that his usual, piecemeal approach to generating and placing his works was safer, particularly during the Depression. He did have several comfortable ongoing commissions such as "Brutus" and his "Raggedy Ann" illustrated proverbs that appeared daily in newspapers across the country.

On the merchandising front, Gruelle had authorized Needlecrafters of Babylon, Long Island, to produce Raggedy pillows and bedspreads, and E. E. Fairchild Corporation of Rochester, New York, to sell playing cards based on his designs. He continued pitching new book ideas to juvenile publishers—among them, Harper Bros., Donohue (reprinter of his older Raggedy Ann titles), and Whitman. And he was working on designs for a new set of dolls based on his current, more stylized book illustrations in hopes of getting a new manufacturer interested. Eventually, however, Gruelle could see that delays and failed deals were outnumbering signed contracts and actual merchandise.

In the fall of 1934, Gruelle summoned help, hiring well-known agent Fred A. Wish. Knowing that Wish was accustomed to representing high-profile artists and authors such as Howard Garis and Fontaine Fox (and handling licensing for such popular characters as Uncle Wiggily, Skippy, and Toonerville Trolley), Gruelle felt the need to assure him of his own creativity and output.

"I like to work," Gruelle wrote, "and never have to scramble around for ideas. I work very fast, never taking a full day to write any of the books I have published so far, with the exception of 'The Magical Land of Noom,' which I dictated to a stenographer in three afternoon sessions."

A 1920s Volland Raggedy Andy doll and friend. (Barbara Lauver)

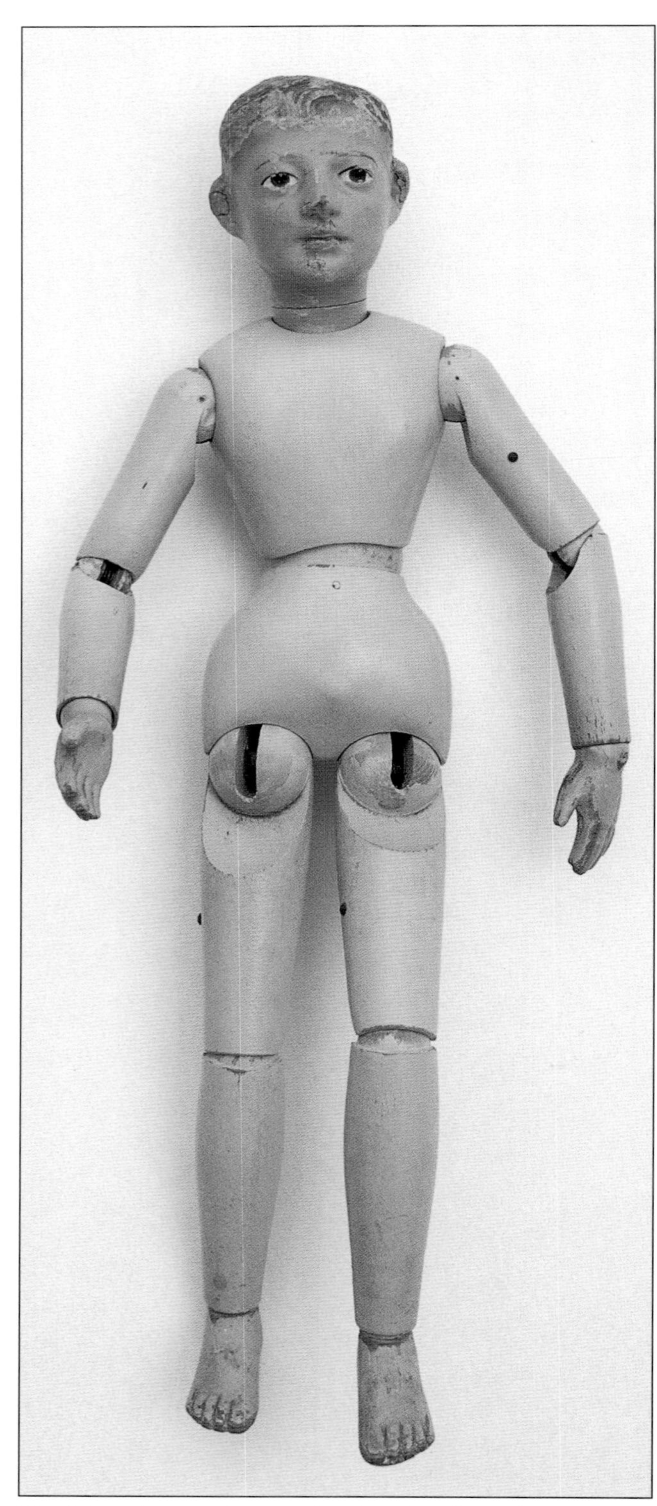

The Schoenhut artist's mannikin that Gruelle used as his model for figure drawing. (Archives of the Miami-Dade County Library System)

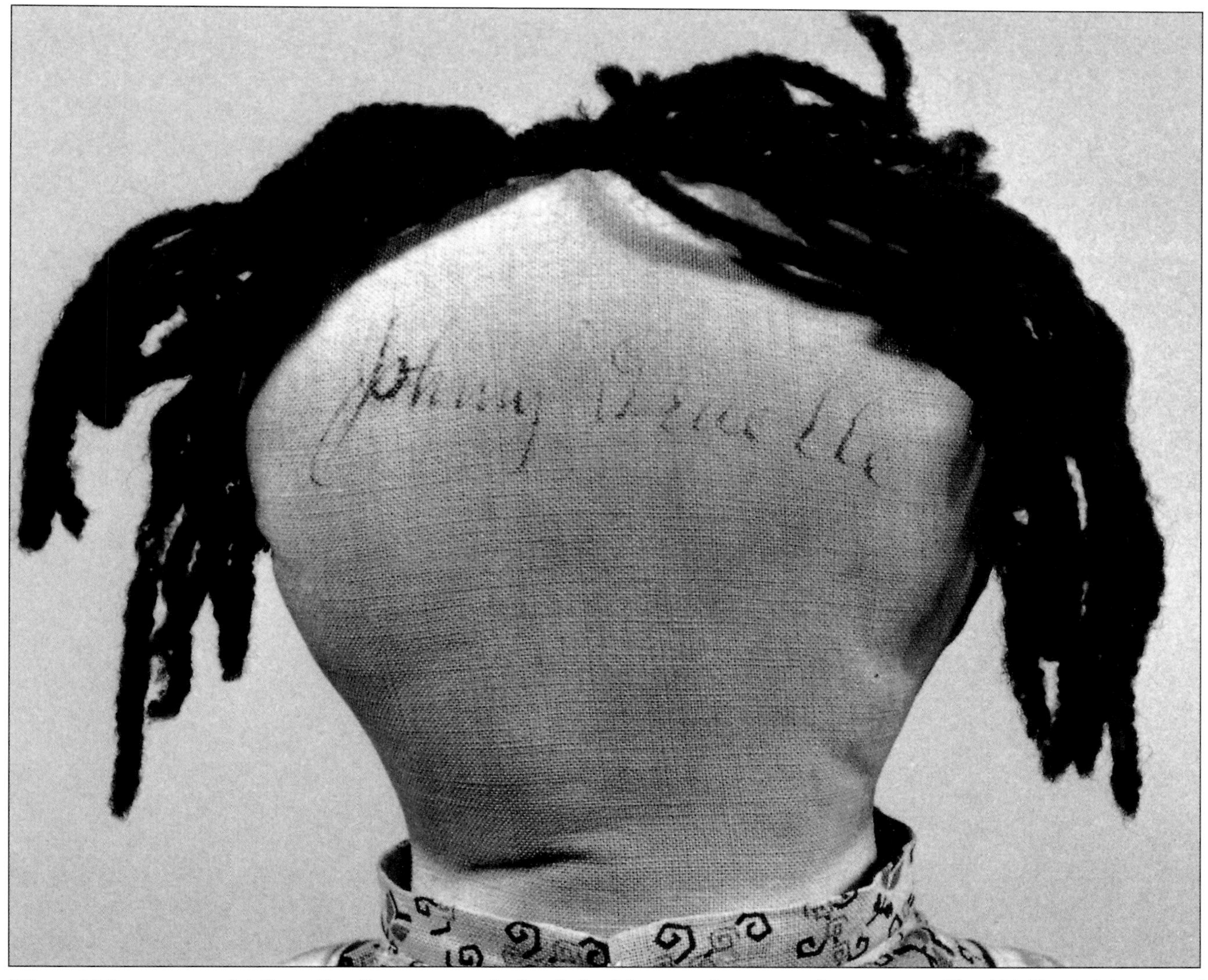

When children presented dolls to Gruelle to autograph, he gladly did so on whatever surface was signable. (Children's Museum of Indianapolis)

> ***One of the best proofs of the salability of these dolls is that the larger department stores in all sections of the country are buying them and are planning to give the dolls a prominent display.***
>
> **Volland Company bulletin**
> **1931**

Wish readily agreed to represent the Raggedys and related characters for dolls, merchandise, motion pictures, radio, and advertising (excluded were Gruelle's book, magazine, and newspaper commissions, which he would continue to oversee directly). Wish was soon negotiating with NBC's Blue Network for a series of 15-minute Raggedy Ann radio shows, while at the same time entertaining an offer from CBS for radio rights to the use of Gruelle's characters.

Confident that Wish could light a fire under his Raggedys, Gruelle felt free to pursue what he loved best—his artwork. Inspired by his south-Florida surroundings, Gruelle quickly wrote and illustrated three new books: *Raggedy Ann in the Golden Meadow*, *Raggedy Ann and the Left-Handed Safety Pin*, and *Raggedy Ann Cut-Out Paper Doll*, each featuring brilliant color illustrations. These new, more streamlined Raggedys were accepted by Whitman for 1935 publication.

THE WHITMAN BOOKS

In 1933, Johnny Gruelle approached Whitman Publishing Company of Racine, Wisconsin, with his well-organized stash of completed Raggedy Ann and Andy manuscripts. Well known among the low-priced children's book publishers that had sprung up during the Great Depression, Whitman (a division of the much larger Western Printing and Lithographing Company) agreed to publish three new Raggedy Ann titles by Gruelle.

Raggedy Ann and the Left-Handed Safety Pin and a paper-doll booklet entitled *Raggedy Ann Cut-Out Paper Doll* were issued in June 1935. During the first six months, sales of both books (Gruelle's first new Raggedy Ann titles in nearly three years) totalled 10,000 copies. In December 1935, the company issued *Raggedy Ann in the Golden Meadow*, an oversized storybook based on newspaper serials for which Gruelle had prepared a new set of dazzling black-and-white and full-color illustrations.

Gruelle continued submitting books and book ideas to Whitman (among them, *Raggedy Ann's Kindness* and *Raggedy Ann and the Queen*), but the company would not publish any more Gruelle titles.

(120)

The Gingerbread Man *(1930), featuring Gruelle's recolored illustrations, was a reprint of the first chapter of Josephine Lawrence's* Man in the Moon Stories *(1922).*

(122)

The lushness and clarity of Gruelle's illustrations for Raggedy Ann in the Golden Meadow *(1935) rank them among his best artwork.*

(123)

(121)

Raggedy Ann and The Left Handed Safety Pin *(1935) reprised a magical motif Gruelle had used in several magazine stories.*

(124)

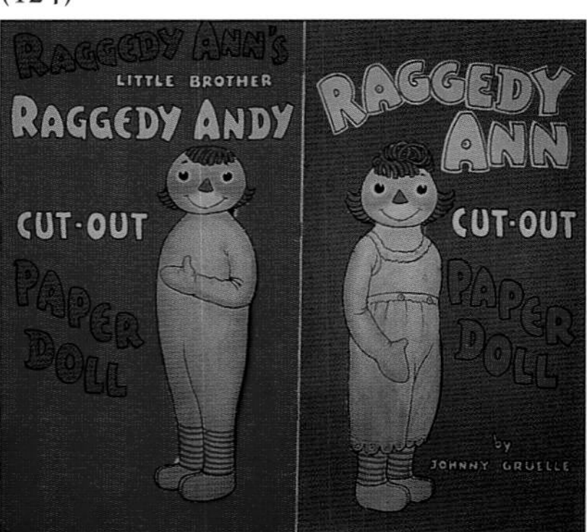

Featuring original verses as well as artwork, Raggedy Ann Cut-out Paper Doll *(1935) is the only published paper doll drawn by Johnny Gruelle. (Candy Brainard/Author's Collection)*

Gruelle's dummy prototype for the never-published Raggedy Ann's Kindness. *(Worth and Suzanne Gruelle)*

Fred Wish continued working on merchandising ideas, alerting advertising agencies of Gruelle's availability as a contract artist. Eager for the work, Gruelle advised his agent, "Should you at any time wish to submit an idea if you will drop me details of what you have in mind; and by that I mean, the type of advertising product, 'tooth paste,' 'chewing gum,' 'breakfast food,' or what-not, I will make the sketches which you may submit to the advertiser."

The fact remained that little if any of Gruelle's work was ever picked up by advertisers. And though Wish continued to assure his client that "no one can do this type of thing better than you can," nothing of note materialized on the licensing front either. But Gruelle and Wish were focusing most of their attention on a different project—getting an authorized Raggedy Ann doll back in production and on the market.

By late fall 1934, negotiations with Exposition Doll and Toy Manufacturing Company of New York City were nearly complete for a brand-new Raggedy Ann doll. Though Wish had received word that some kind of an unauthorized trademarked Raggedy Ann doll was being sold in Kresge department stores, when Kresge officials denied the claim Gruelle's agent did not investigate further. However, in April 1935 a set of unlicensed Raggedy Ann and Andy dolls were exhibited at the New York Toy Fair. They had been manufactured by Molly-'Es Doll Outfitters of Philadelphia, a company known for its elaborate doll costumes and accessories. Gruelle and Exposition were outraged. Their responses set in motion *Gruelle v. Goldman*, a costly and acrimonious trademark infringement lawsuit.

> ***There isn't any question in my mind that a Raggedy Ann show would go big—and if I had not had to plunk down cash for this shanty, and had not stopped myself, I would have started a show myself.***
>
> **Johnny Gruelle to Howard Cox**
> **1933**

Lawsuits over the ownership of character dolls were not uncommon. *Rose O'Neill v. Borgfeldt* and *Richard Outcault v. New York Herald* (cases pertaining to the Kewpie and Buster Brown characters) were but two of many disputes in which creators had sued for the rights to their creations. In *Gruelle v. Goldman*, Gruelle claimed that he had never authorized

Mr.Clayton Cooper,
Committee Of One Hundred,
Miami Beach,Fla.

Dear Mr.Cooper:

I wish to thank you and all the members of The Committe Of One Hundred for the honor you do me.

It is certainly a great privilege and pleasure to be associated with such a wonderful organization.And,aside from putting a crop of whis--kers on the members and,size forty eights on the forms of their wives,I shall try not to bring dis-credit to the Club.

Gratefuly yours,

Johnny Gruelle

4455 Meridian Ave.
February ten,
Nineteen thirty three.

Shortly after he made Florida his home, Johnny Gruelle was voted into the exclusive Committee of One Hundred of Miami Beach. His illustrated letter of acceptance was characteristically (and humorously) humble.

A STORYBOOK DOLL: THE EXPOSITION RAGGEDY ANN

When the wide-eyed Exposition Raggedy Ann doll made her debut in 1935, she was a close match for the graceful, more streamlined rag-doll characters that Gruelle had been drawing for several years. Her silk-screened features give her a smooth, open expression. Rather than traditional shoe buttons, her eyes are part of the screened design. Her lush wig is bright burgundy-colored wool yarn, styled with a prominent topknot.

At 18" tall, the lanky Exposition Raggedy Ann was sewn and stuffed so that she could sit up, her-red-and-white-striped legs extending up to her bottom. She wears removable black felt shoes, a light-blue-pink-and-white floral print dress, and a white see-through apron and pantalettes. Her clothing is edged either in white or light-blue binding. She has no heart, but a triangle-shaped paper hang tag and a sewn-on cloth label on the lower edge of her dress clearly identify her as Johnny Gruelle's authorized creation.

Unfortunately, the Exposition Raggedy Ann remained in production only until 1936, the company's plans to sustain its license (and introduce a matching Raggedy Andy doll) dashed by the protracted *Gruelle v. Goldman* lawsuit.

The Exposition doll personified the wide-faced, topknotted Raggedy Anns that Gruelle was rendering during the 1930s. (Candy Brainard/ Author's Collection)

(125)

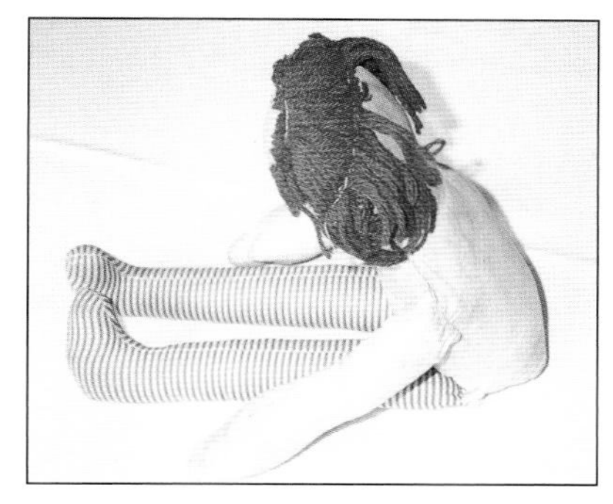

The Exposition Raggedy Ann, whose long, striped legs had seams up the front, was designed to sit up. This example was signed by Johnny Gruelle in 1935. (Bernette Benedict)

(126)

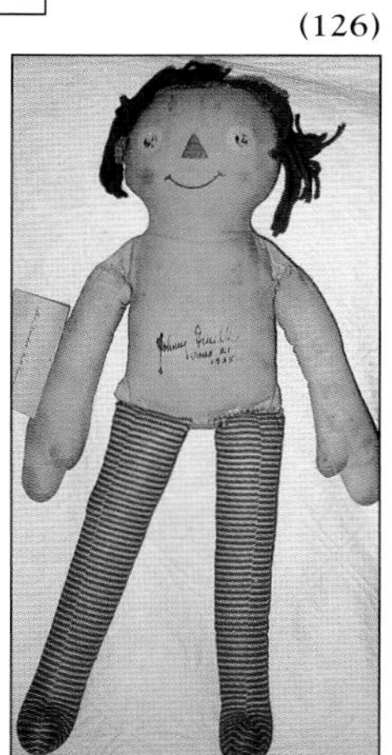

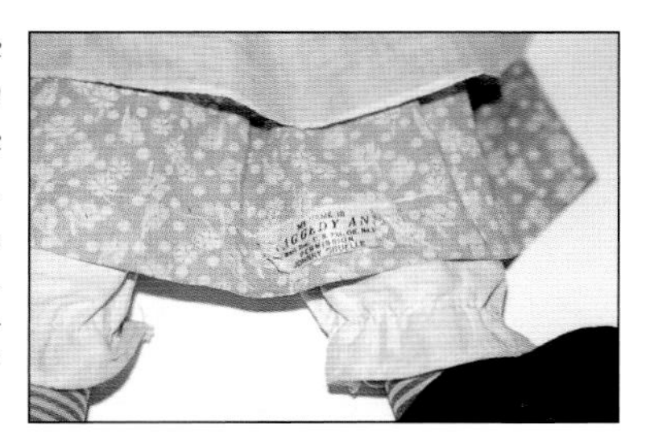

The cloth label on the Exposition Raggedy Ann read: "My Name is Raggedy Ann. Trade Mark Reg. U.S. Patent Office. No. 107,328. Permission Johnny Gruelle." (Candy Brainard)

(127)

Exhibit "M" in the Gruelle v. Goldman *lawsuit. (National Archives)*

his Raggedy Ann and Andy dolls to be manufactured and sold by Molly-'Es Doll Outfitters; that he had not abandoned his Raggedy Ann trademark as Goldman claimed he had; and that the design patent Molly-'Es had registered for a Raggedy Andy doll should be declared invalid. (For a more detailed account of *Gruelle v. Goldman,* the reader is directed to the illustrated biography *Johnny Gruelle, Creator of Raggedy Ann and Andy.*)

For the next several months, Gruelle, his attorney John W. Thompson, and what was left of the staff at the downsized Gerlach-Barklow Company scurried to reconstruct the manufacturing history of Raggedy Ann and Andy to present as part of their court defense. While kicking themselves for not having launched a Gruelle family-owned corporation, Gruelle and his attorney built their legal case around the fact that the trademarked name "Raggedy Ann" had been applied by Volland not only to dolls but also to other merchandise such as bookshelves and that the name was not merely a generic designation for dolls.

Thompson sternly warned Goldman's attorneys that "careful consideration should be given to distinguishing between that which may have become a part of the public domain by reason of the expiration of Gruelle's design patent and that which has come to be associated in the mind of the public with the named 'Raggedy Ann.'"

Meanwhile, driven by a belief that Gruelle's Raggedys were, in fact, hers to do with what she wished, Mollye Goldman continued manufacturing and selling Raggedy Ann and Andy dolls, intimidating major retailers into buying only those made by her company. She widely advertised her Raggedys, made them available as magazine subscription premiums, and provided them to screen stars such as Virginia Weidler to pose with in publicity photographs. In 1935, a Molly-'Es Raggedy Andy even made its way to singer and radio personality Kate Smith, who had appealed over the air for listeners to contribute dolls for poor children. It seemed that the Molly-'Es Raggedys were everywhere.

<John> Thompson and I are really going after things now, and something is certain to pop on R. Ann.

Johnny Gruelle to Howard Cox
July 1934

An important but unanswered question is whether Gruelle and Goldman had *ever* considered working together. Preliminary discussions could have taken place in 1934 when Volland was about to go out of the doll-purveying business and Gruelle was searching for a new manufacturer. Talks also could have occurred later, perhaps during the lawsuit. In April 1936, a mystifying full-page advertisement appeared in *Playthings,* announcing (just in time for the New York Toy Fair) that Goldman and Gruelle were in partnership.

Whether this indicated a nascent deal or was simply an attempt to paint a dismal picture with a rosy brush is not known. In her later years, Mollye Goldman claimed she still had in her file of contracts some kind of 1936 agreement with Johnny Gruelle; however, no sort of agreement, informal or written, was ever mentioned by either side in sworn affidavits.

It seems as if everybody wants it except someone who is willing to pay a price. The dolls are copied, the names applied to every conceivable kind of product, people from coast to coast want to produce the plays, salads are named after "Raggedy Ann," etc. Apparently everybody knows it and likes it except somebody to whom a legitimate use of the property could be made to bring a return in goodwill. What is wrong with this picture?

John Thompson to Fred Wish
November 1935

Gruelle v. Goldman was decided in July 1936 in Mollye Goldman's favor. The judge's ruling, issued on October 5, 1936, stated that Gruelle had, in fact, abandoned his Raggedy Ann and Andy trademark for a period of time. Gruelle's attorney immediately filed an appeal, and in December 1937 Johnny finally won his lawsuit. A deciding factor was that production of the Molly-'Es Raggedys had actually begun in November 1934—a month before Volland had assigned the Raggedy rights and trademark back to Gruelle, making Goldman guilty of trademark infringement after all.

Johnny Gruelle, 1934.

ENDEARING INFRINGERS: THE MOLLY-'ES RAGGEDY ANN AND ANDY DOLLS

There is no denying that the Molly-'Es Raggedy Anns and Andys are lovable, engaging, and attractive dolls. Deftly designed and skillfully constructed, they stand as an important evolutionary link between the more homespun, heterogeneous dolls sold by the P. F. Volland Company and the sleeker, more standardized dolls produced by Georgene Novelties. Though unauthorized, and later declared infringements, the Molly-'Es Anns and Andys are, in quality and design, on a par with the best of the licensed Raggedys.

(128)

(129)

Although the Molly-'Es Raggedy Ann and Andy dolls were not advertised until 1935, production actually began in November 1934. (Barbara Dubay. Photos by Craig McNab)

(130)

A rare Molly-'Es Raggedy Andy with button eyes. (Jacki Payne)

(131)

(132)

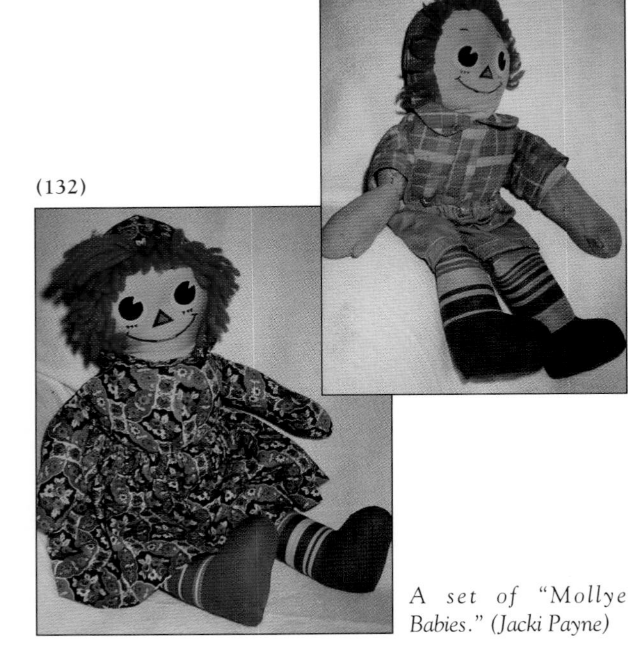

A set of "Mollye Babies." (Jacki Payne)

(133)

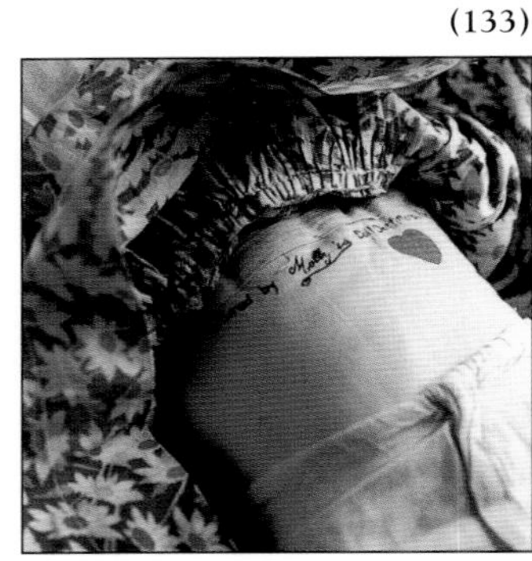

The chest fabric of some Molly-'Es dolls was stitched together or taped over to obscure the names Raggedy Ann and Raggedy Andy. Eventually reduced to a simple, solid red heart, the slowly disappearing chest logo most likely resulted from the escalating Gruelle v. Goldman *lawsuit. (Bill Miller and Alison Hubbard)*

(Note: Doll identifiers that appear in quotation marks are either commonly used terms or the author's nomenclature—not official company designations. They are presented here to assist in identifying and classifying representative Molly-'Es dolls.)

"Archetype" Raggedy Ann and Raggedy Andys (1935-37). Except for their clothing, the Molly-'Es "Archetype" Raggedy Anns and Andys (so dubbed because of their prototypical appearance) are essentially the same doll. Produced in both an 18" and slightly larger 22" size, these dolls have boldly silk-screened faces (whose designs evolved somewhat over time) incorporating noses outlined in black, curved, black smile lines with straight red centers, and large "pie" eyes (notched black circles set to the left inside larger white circles) set off by thin black eyebrows and three black triangular lower lashes. Legs are usually of blue-red-yellow-white, multi-striped fabric and sewn-on feet, of blue fabric. The "Archetype" dolls are wigged with topknots in reddish-orange yarn.

The "Archetype" Raggedy Anns wear dresses of cotton (most often a large daisy print) and white organdy aprons with ruffles and pantaloons with eyelet trim. The "Archetype" Raggedy Andys are dressed in pants sewn to a geometric print shirt, silk ties, and circular blue caps with wide white brims. The "Archetype" dolls came with red or blue paper wrist tags and were packaged for sale in their own boxes.

Mollye Goldman's advertising for the dolls she touted as hers often included her invalidated patent design for Raggedy Andy.

During the fall of 1935, Curtis Publishing Company offered a set of Molly-'Es Raggedys as subscription premiums for Country Gentleman *and* Ladies' Home Journal.

During three years of production, a number of "Archetype" Raggedys appeared with noticeable anomalies—dolls with button, rather than painted-on eyes; dolls lacking nose outlines or eyelashes; dolls whose facial features were not in register; and dolls made with two faces sewn together to form the head.

"Mollye Babies" (ca. 1935-37). Molly-'Es Doll Outfitters also produced a set of 14" dolls, dubbed "Mollye Babies" by collectors. These cherubic dolls have multi-striped legs, blue fabric feet, print torsos, and fabric "caps" forming the backs of their heads, with a fringe of yarn "hair" sewn in at the hairline. Instead of real shirts, they have print arms (or flesh-colored ones with sewn-on print sleeves) and sewn-on clothes; only Ann's skirts, pinafores, and panties, and Andy's trousers are removable. The "Mollye Babies'" silk-screened faces consist of a wide linear smile, an outlined red nose, and wide "saucer" eyes with small pie-shaped slivers cut out. Some "Mollye Babies" were made with a bottom they could sit on; others were not.

"Beloved Belindy" (only prototype documented). Mollye Goldman claimed that she also produced a Beloved Belindy doll that was sold in limited quantities in several stores before the *Gruelle v. Goldman* lawsuit forced them off the market. However, nothing beyond a prototype has been documented.

(134)

(135)

This pair of 1935 Molly-'Es dolls were presented as plaintiff's exhibits in the Gruelle v. Goldman *lawsuit. (National Archives)*

NOT QUITE RAGGEDY: THE SPECTRE OF UNAUTHORIZED MERCHANDISE

During Johnny Gruelle's lifetime and even afterward, there were repeated attempts by businesses and individuals to market unlicensed Raggedy Ann and Andy merchandise. Driven partly by a belief that Gruelle's Raggedys were (or should be) public domain characters and partly by greed, offenders would use unauthorized images of Raggedy Ann and Andy in their products or advertising without securing permission or a license. The most notorious infringement occurred in the 1930s when unauthorized Raggedy Ann and Andy dolls were produced by Molly-'Es Doll Outfitters.

Some companies requested (and assumed they should have) free use of the characters in everything from toys to salads; sewing patterns to wrapping paper; dolls to puppet plays. Others sought no permission, hoping to avoid the cost of a license by not naming their clearly derivative products "Raggedy Ann" or "Raggedy Andy," or by disguising one or a few of the Raggedys' key characteristics while still capitalizing on the dolls' recognition value.

The violators certainly should have known better. Gruelle's copyrights and trademarks were clearly indicated on authorized works and merchandise, and the Johnny Gruelle Company was an active and visible presence in the playthings marketplace. With each

(140)

(141)

Unauthorized greeting cards, ca. 1940s-1950s. (Evon Webster/Author's Collection)

(136)

(137)

(138)

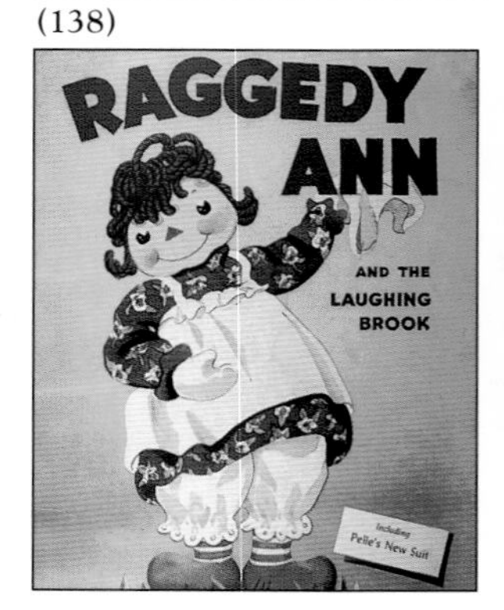

(139)

These Mary Perks booklets were yet another unlicensed repackaging of the McLoughlin Bros.' Raggedy Ann series.

(142) (143) (144)

The American Crayon Company issued unauthorized reprints of the McLoughlin Bros.' "Little Color Classics," with illustrations by the Stovers clearly modeled after Justin Gruelle's artwork. A legal challenge by the Johnny Gruelle Company in the mid-1940s seemed not to dampen brisk sales of these booklets.

infringement, Gruelle and his family (and later, the Johnny Gruelle Company) grew more wary and protective of their intellectual property, having experienced, firsthand, the damage that unlicensed activity could cause.

Gruelle's agent, Fred Wish, perhaps put it best in his affidavit for the *Gruelle v. Goldman* lawsuit. Speaking about the consequences of unauthorized uses of copyrighted and trademarked characters, he noted:

> After many years of experience . . . I am convinced that the public . . . is susceptible to the appeal of . . . characters when their names, or slogans or expressions . . . are attached with the sale of commodities other than the . . . works in which the characters originated or gained their public following. The unauthorized use of popularized names and characters tends to destroy the value . . . and diverts the reward from the creator to those who have done nothing toward building the popularity and goodwill associated with the names in question.

(146)

(145)

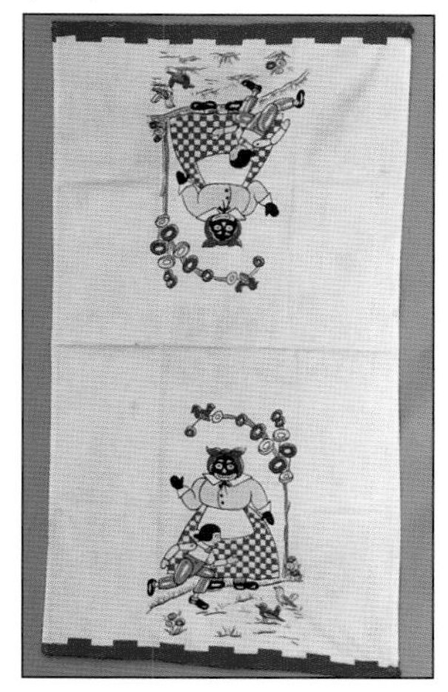

Other apparently unlicensed merchandise from the 1940s and 1950s. (Candy Brainard/ Brenda Milliren)

The "Raggedys" are peaceable folks stuffed with soft wadding but they "pack a big wallop" for those who threaten to hurt them or me.

Johnny Gruelle
Playthings
May 1935

Molly-'Es Doll Outfitters was ordered to pay Gruelle damages and prohibited from any further manufacture of Raggedy Ann and Andy dolls. The appellate court further noted that *Gruelle v. Goldman* could have been much less protracted had Gruelle and his attorney sued on grounds of copyright rather than trademark infringement.

The Molly-'Es dolls would not be the last infringement Raggedy Ann and Andy would suffer, but it was certainly the most damaging. Though he won his lawsuit, Gruelle was faced, once again, with finding a manufacturer to produce his dolls since Exposition, because of the legal dispute, had ceased production of its Raggedy Ann dolls and abandoned its plans for a Raggedy Andy. During the course of the lawsuit, Gruelle's health had deteriorated markedly. He was seeing a doctor for an ongoing heart condition, he suffered from chronic pain, and he was emotionally worn down. On January 9, 1938, only three weeks after *Gruelle v. Goldman* was settled in his favor, Johnny Gruelle died of a heart attack.

As she mourned her husband and sweetheart, Myrtle Gruelle also took stock of the grim business situation. She knew it was up to her and her family to ensure the future of Raggedy Ann and Andy. Her attention was diverted briefly in the spring of 1938 when Mollye Goldman, threatened by bankruptcy, again asserted her ownership of the Raggedys, appealing *Gruelle v. Goldman* to the U.S. Supreme Court. Although the judges denied Goldman's request for a *writ of certiorari*, finally laying the case to rest, Goldman would never fully pay off the court-ordered damages awarded to the Gruelle family.

The Molly-Es business has given us quite a set back with our new connection to the dolls. It is unfortunate that we did not complete the incorporation of your company last fall when we considered it because that would have headed off this particular piece of piracy.

Attorney John Thompson to Johnny Gruelle
June 1935

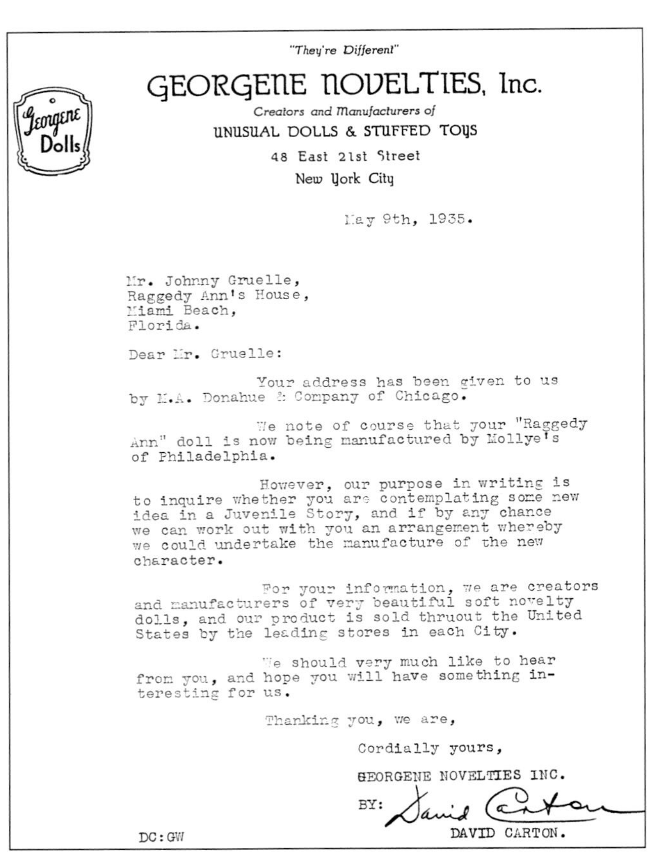

"They're Different"

GEORGENE NOVELTIES, Inc.

Creators and Manufacturers of

UNUSUAL DOLLS & STUFFED TOYS

48 East 21st Street

New York City

Georgene Dolls

May 9th, 1935.

Mr. Johnny Gruelle,
Raggedy Ann's House,
Miami Beach,
Florida.

Dear Mr. Gruelle:

Your address has been given to us by M.A. Donahue & Company of Chicago.

We note of course that your "Raggedy Ann" doll is now being manufactured by Mollye's of Philadelphia.

However, our purpose in writing is to inquire whether you are contemplating some new idea in a Juvenile Story, and if by any chance we can work out with you an arrangement whereby we could undertake the manufacture of the new character.

For your information, we are creators and manufacturers of very beautiful soft novelty dolls, and our product is sold thruout the United States by the leading stores in each City.

We should very much like to hear from you, and hope you will have something interesting for us.

Thanking you, we are,

Cordially yours,

GEORGENE NOVELTIES INC.

BY: David Carton

DAVID CARTON.

DC:GW

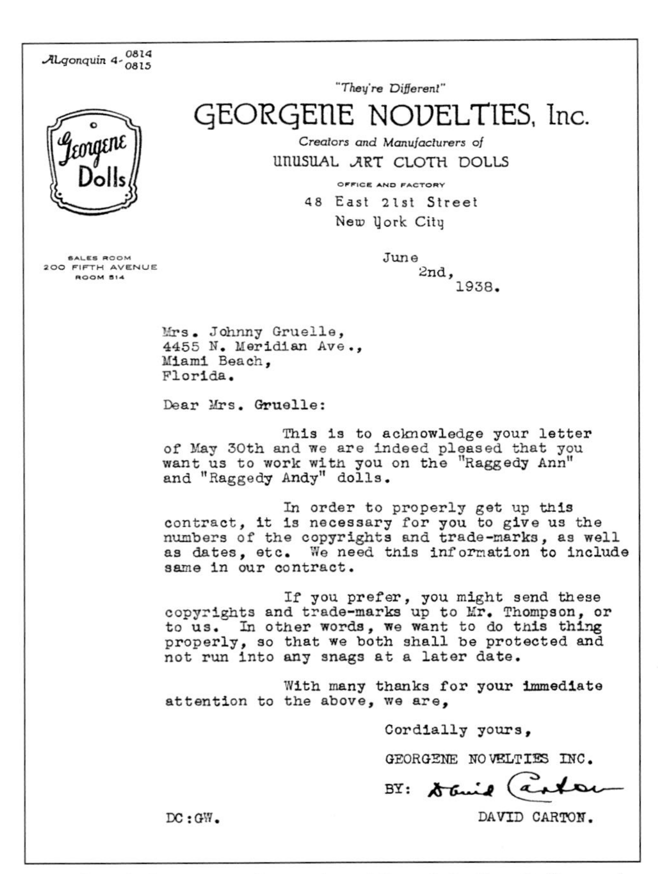

ALgonquin 4-0814 0815

"They're Different"

GEORGENE NOVELTIES, Inc.

Creators and Manufacturers of

UNUSUAL ART CLOTH DOLLS

OFFICE AND FACTORY

48 East 21st Street

New York City

Georgene Dolls

SALES ROOM
200 FIFTH AVENUE
ROOM 514

June 2nd, 1938.

Mrs. Johnny Gruelle,
4455 N. Meridian Ave.,
Miami Beach,
Florida.

Dear Mrs. Gruelle:

This is to acknowledge your letter of May 30th and we are indeed pleased that you want us to work with you on the "Raggedy Ann" and "Raggedy Andy" dolls.

In order to properly get up this contract, it is necessary for you to give us the numbers of the copyrights and trade-marks, as well as dates, etc. We need this information to include same in our contract.

If you prefer, you might send these copyrights and trade-marks up to Mr. Thompson, or to us. In other words, we want to do this thing properly, so that we both shall be protected and not run into any snags at a later date.

With many thanks for your immediate attention to the above, we are,

Cordially yours,

GEORGENE NOVELTIES INC.

BY: David Carton

DAVID CARTON.

DC:GW.

David Carton's letters to Johnny and Myrtle Gruelle would initiate a two-decade business relationship. (Cox-Scheffey Collection)

One of the chief reasons for Mrs. Gruelle's keen desire to organize the partnership known as the Johnny Gruelle Company was her dissatisfaction with the financial results of Donohue's effort with the Raggedy Ann property.

Howard Cox to Robert Dietz
January 1942

Myrtle Gruelle forged ahead, eventually coming across a letter sent to Johnny in 1935 by David Carton, president of a New York doll manufacturer. Carton had written to express interest in producing dolls based on new characters Gruelle might be creating for forthcoming books. Though the query had piqued her late husband's interest, the escalating legal problems with Molly-'Es had forestalled any developments. However, the time now seemed right, and by spring of 1938 Myrtle Gruelle had connected with Carton and his company, Georgene Novelties.

Bearing the name of founder Georgene Hopf Averill, Georgene Novelties specialized in unusual dolls and stuffed toys that were sold to department stores throughout the U.S. Though Madame Averill (as she was known) and her husband, Paul, maintained a link with the company named after her, as of the late 1930s she had handed over Georgene Novelties to her brother Rudolph Hopf (who handled manufacturing) and David Carton.

With new designs approved and potential retailers lined up, on November 23, 1938, Myrtle Gruelle authorized Georgene Novelties to begin manufacturing a brand-new set of Raggedy Ann and Raggedy Andy dolls, soon to be followed by a newly designed Beloved Belindy.

Myrtle focused next on merchandising, turning to Howard Cox for advice on how to handle her late husband's characters. Cox later described that meeting, recalling that Myrtle had been "beside herself in an effort to restore to a reasonable degree the earning power of the Raggedy Ann property."

BUDDY AND SIS:
THE HUGABLE NURSERY PETS

Sometime during the late 1930s or early 1940s, the Philadelphia-based American Toy & Novelty Manufacturing Company introduced two new dolls in its Hugable <*sic*> Nursery Pet line. Well known for its cloth-bodied dolls dressed as cowboys, soldiers, and other characters, American Toy & Novelty had apparently not secured authorization or a license to produce Buddy and Sis, each of which bore an uncanny resemblance to Raggedy Ann and Andy.

The Buddy and Sis dolls measure approximately 15" and are constructed of cotton muslin stuffed with kapok. Their clothing, which varies widely from doll to doll, is an a exuberant mix of materials sewn together, creating a crazy-quilt, "ragamuffin" look. Fabrics (also used for the dolls' legs, torsos, and arms) include cotton chambray, muslin, broadcloth, taffeta, and eyelet, in prints ranging from solids and stripes to plaids and florals.

(147)

(148)

(149)

The Hugable Nursery Pets were dressed in a multitude of bright "rag-bag" fabrics. (Jacki Payne/Author's Collection/Brenda Milliren)

Sis wears a blouse (sleeves sewn to a matching fabric torso), removable skirt, and pantalettes. Buddy wears a shirt (sleeves sewn to the torso), removable pants, and sailor-style cap. Feet are most often rayon, but other fabrics were also used. The dolls were issued with a sewn-on cloth label identifying them as either "Sis" or "Buddy," and each came packaged in its own box.

Most notable about the Hugable Nursery Pets are their shiny oil-cloth faces. Printed facial features (which evolved over time) include medium-to-large white eye circles overprinted with smaller slotted black circles, small straight black eyebrows, and three or four small lower lashes. Mouths are two arc-shaped lines set on either side of an elliptical red center. The small, red triangular noses (with either straight or slightly curving sides) are outlined in black, with printed "stitching." Hair is made from various shades of reddish-brown yarn.

American Toy & Novelty may have thought it was legally in the clear by not calling its dolls Raggedy Ann or Raggedy Andy and by changing the traditional Raggedy design slightly. Buddy and Sis were but two of many look-alike infringement Raggedy dolls (most of them unmarked, a number of them with oil-cloth faces) produced during this time.

The resemblance of the Hugable Nursery Pets' face design to the discontinued Molly-'Es dolls of the mid-1930s (and the fact that both companies were based in Philadelphia) has led to speculation that Mollye Goldman may have had something to do with the creation and/or marketing of Buddy and Sis, which were produced and sold as late as the 1950s.

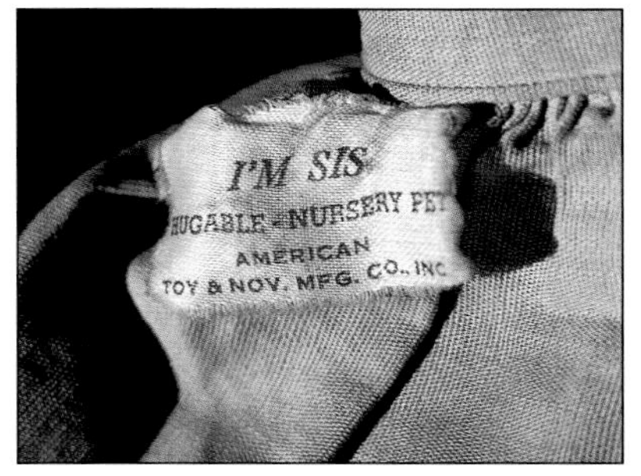

Although tags and boxes gave no indication of manufacturing dates, most collectors agree that on the basis of fabrics and periods of availability, the Hugable Nursery Pets were produced between the late 1930s and early 1950s.

AFTER JOHNNY'S HEART: THE GEORGENE RAGGEDY ANN AND ANDY DOLLS

The sweet-faced dolls manufactured by Georgene Novelties rank among the most memorable of all the commercial Raggedy Anns and Andys. Perhaps because they, more than any others, capture the essence and whimsy Gruelle had in mind when he first conceived of the Raggedys, the Georgene dolls are, time and again, voted a collectors' favorite.

However, authenticating their two-decade production history is another story. The process is complicated by the lengthy license period, sparse company records, and ongoing confusion concerning the multiple dates appearing on the dolls' cloth body tags—dates that reference book copyrights rather than actual doll manufacture dates. However, using what primary and secondary source material is available, as well as provenance information, the Georgene Raggedy Ann and Andy dolls can be organized into several historical groupings.

(150)

(151)

The inaugural "Outlined Nose" Georgene Raggedys were clean, precise interpretations of Gruelle's rag doll design. (Pauline Baker)

(Note: Doll identifiers that appear in quotation marks are either commonly used terms or the author's nomenclature—not official company designations. They are presented here to assist in identifying and classifying representative Georgene dolls.)

"Outlined Nose" Raggedy Anns and Raggedy Andys (1938-early 1940s). The inaugural Georgene Novelties Raggedy Ann and Andy dolls, first released in late 1938, stand approximately 19" tall. The bodies of these meticulously manufactured dolls (Ann and Andy were cut from the same pattern) are made of cotton stuffed with lightweight kapok. Faces are stamped with curved smile lines with red centers, white eye circles (with black metal eyes affixed in the center), four lower lashes, linear black eyebrows, two small red pinpoints between the eyes, subtle cheek blushing, and well-proportioned noses boldly outlined in black. The latter feature is so distinctive that it has inspired the collector-coined identifier "Outlined Nose."

During the early 1940s, Georgene Raggedys replaced the Molly-'Es dolls as Curtis Publishing Company subscription premiums.

This is the famous pair, 19 inches tall. Yarn hair, on and off clothes, irresistible painted faces with button eyes. These are not ordinary dolls.
656. Raggedy Ann. $2.25
657. Raggedy Andy. $2.25

The "Outlined Nose" Georgene Raggedy Ann and Andy were offered in the September 1941 Robert Kellogg catalogue. (Marge Meisinger)

(152)

By the 1950s, Georgene Novelties was manufacturing 45" Raggedys. (Jacki Payne)

The "Outlined Nose" Raggedy Anns and Andys are wigged with thick, reddish-brown wool yarn reinforced by netting underneath, Ann's wig incorporating a topknot. Raggedy Ann wears a dress of bright, floral-patterned cotton (with small gold safety pins instead of buttons), along with white cotton pinafore and pantalettes. Raggedy Andy wears a shirt of diagonally-printed cotton sewn to solid blue pants (set off with hollow white- painted metal buttons), a dark ribbon tie, and round, blue fabric cap edged in white.

Both "Outlined Nose" dolls have red-and-white-striped legs (several different stripe widths have been documented) and black fabric feet. Each also has an "I Love You" heart stamped on its chest and a printed cloth tag sewn into the left side seam (red for Ann, blue for Andy) and heart-shaped paper hang tag that state trademark and book copyright information.

(153)

A number of "Wartime" Raggedys possessed legs and shoes made from varying printed fabrics. (Jacki Payne)

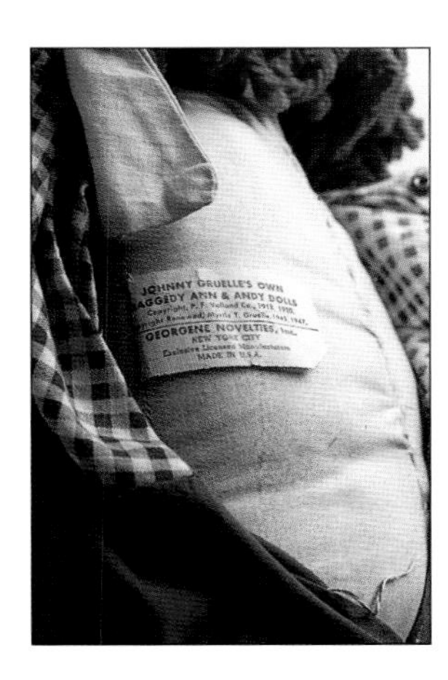

Georgene Raggedys had cloth tags sewn into their left side seams. Although the earliest were printed in red for Raggedy Ann and blue for Raggedy Andy, all tags from the 1940s on were printed in red. (Bill Miller and Alison Hubbard)

"Awake-Asleep" Raggedy Ann and Raggedy Andys (ca. 1940-46). Around 1940 or so, Georgene Novelties introduced a smaller pair of double-faced Raggedy Anns and Andys, referred to by collectors as the "Awake-Asleep" dolls. Measuring 12-13", these dolls have a fringe of reddish-brown yarn "hair" sewn in at the head seam. Facial features are stamped, noses are outlined in black, and eyes on the awake side are painted metal discs held by prongs. Ann's apron is continuous, with a seam at the side. The red-and-white-striped legs of the "Awake-Asleep" dolls extend slightly above the waist.

Around 1944, the "Awake-Asleep" Raggedy Anns and Andys were redesigned, losing their nose outlines in the process.

continued on next page

(154) (155)

"Awake" and "Asleep" sides of the early (left) and later (right) Georgene "Awake-Asleep" Raggedys. (Author's Collection/ Candy Brainard)

(156)

The Georgene Raggedys were sold in labeled boxes and wore heart-shaped paper hang tags identifying them. (Brenda Milliren)

"Wartime" Raggedy Anns and Raggedy Andys (ca. 1943-46). By 1943 or so, Georgene was producing newly designed 23" and 31" Raggedys in addition to its standard 19" set—dolls designated here as the "Wartime" Raggedys. Noses of the "Wartime" dolls lost their familiar black outline and took on slightly curved sides. Materials shortages resulted in "Wartime" Raggedy Anns and Andys with wigs of cotton rather than wool yarn, fewer metal buttons on clothing, and a wider variety of fabrics (probably whatever was available) for dolls' dresses, caps, legs, and feet.

"Postwar/Silsby" Raggedy Anns and Raggedy Andys (1946-ca. 1950). By 1946, the Georgene "Awake-Asleep" dolls were being phased out, and the company's 19", 23", and 31" dolls had once again been redesigned, accompanied by new 15" dolls. Most of these dolls, which appeared following the end of World War II, were issued with Myrtle Gruelle's new married name "Silsby" printed on their cloth tags and boxes—hence the identifier "Postwar/Silsby" dolls. When Myrtle Gruelle's second marriage ended in the late 1940s, the "Silsby" name eventually disappeared from tags and boxes.

(158)

(157)

A pair of "Fifties" Georgene dolls, with tin eyes and lush, glued-on wigs. (Bill Miller and Alison Hubbard)

"Fifties" Raggedy Anns and Raggedy Andys (ca. 1950-early 1960s). By the early 1950s, Georgene Raggedys were available in sizes ranging from 15" to 45". These dolls fall into an ever-evolving style descended directly from the "Postwar/Silsby" dolls and collectively referred to here as the "Fifties" Raggedy Anns and Andys. Changes include glued-on (rather than sewn-on) wigs and a return to painted metal eyes. As the "Fifties" Raggedys continued evolving during the late 1950s into the early 1960s, their smiles became more and more curvaceous, giving these dolls a saucier, more dimpled look.

Advertising for "Fifties" Georgene Raggedy Ann and Andy dolls. (Marge Meisinger)

(159)

Attempts to accommodate a too-large facial design on smaller Georgene "Fifties" Raggedys resulted in dolls that appear to grimace, rather than smile.

"Plain Tag" Raggedy Anns and Raggedy Andys (ca. 1962). The final Georgene Raggedys, produced in the early 1960s, comprise a mysterious group. These seeming anomalies, referred to here as "Plain Tag" dolls, appear in every other way to be Georgene Raggedys, except for the flesh-colored or white cloth tags made of plain body material sewn into their side seams instead of the standard printed tag. Collectors have theorized that these tags, which bear single initials or hand-stamped names (among them, "Torcia," "Kirget," "Wright," and "Lapenta"), probably represent the names of individuals or supervisors in charge of fabricating or finishing the dolls at a time close to, or immediately following, the expiration of Georgene Novelties' license.

(160)

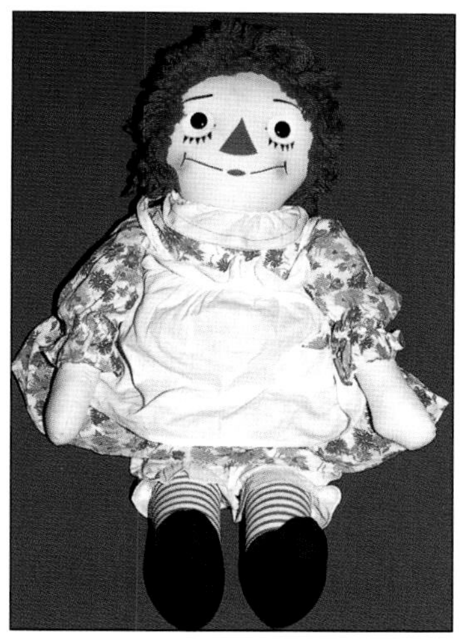

One of the mysterious "Plain Tag" Georgene Raggedys and her hand-stamped tag made of body fabric.

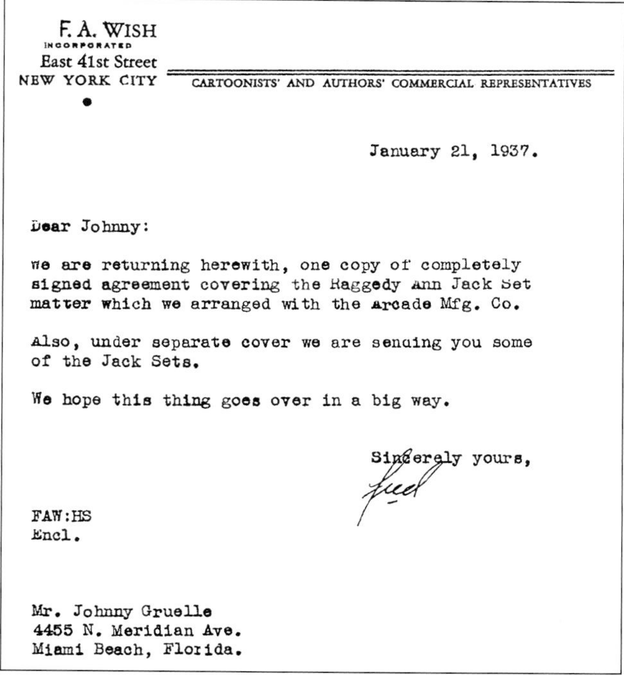

F. A. WISH
INCORPORATED
East 41st Street
NEW YORK CITY

CARTOONISTS' AND AUTHORS' COMMERCIAL REPRESENTATIVES

January 21, 1937.

Dear Johnny:

We are returning herewith, one copy of completely signed agreement covering the Raggedy Ann Jack Set matter which we arranged with the Arcade Mfg. Co.

Also, under separate cover we are sending you some of the Jack Sets.

We hope this thing goes over in a big way.

Sincerely yours,

Fred

FAW:HS
Encl.

Mr. Johnny Gruelle
4455 N. Meridian Ave.
Miami Beach, Florida.

In 1937, Johnny Gruelle copyrighted designs for Raggedy Ann and Andy jacks-and-ball sets produced by Arcade Manufacturing Company of Freeport, Illinois. The sets were among the few licensed products from the 1930s.

Picking up where they had left off several years before, Cox, Myrtle, and the Gruelles' twenty-two-year-old son, Richard, a sharp businessman, began plotting a company that could oversee Raggedy Ann and Andy. On September 1, 1939, the three formed a business called the Johnny Gruelle Company. Chartered as a limited partnership, the new company's stated mission was to publish and reprint Gruelle's works and coordinate the active licensing of Raggedy Ann and Andy merchandise.

Shortly before he died, Gruelle and his agent Fred Wish had worked on several licensing ideas, among them a new set of Raggedy birthday cards, coil-spring dolls (which Gruelle had dubbed "The Jitters Family"), and a possible revival of Gruelle's Quacky Doodles toys. Among the very few products actually authorized and produced during this time, however, were several jacks-and-ball sets in packages decorated with original Gruelle designs. Unfortunately, the *Gruelle v. Goldman* lawsuit, general

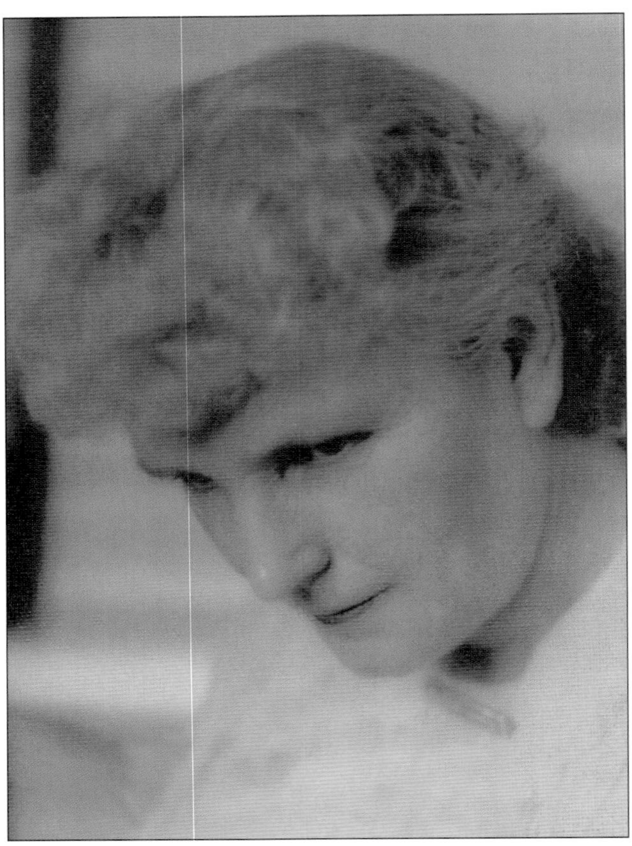

Myrtle Gruelle, Howard Cox, and Richard Gruelle—the three founding partners of the Johnny Gruelle Company. (Tom and Joni Gruelle Wannamaker/Cox-Scheffey Collection/Ruth Gruelle)

Depression conditions, and Wish's seeming inability to secure viable licensees had compromised full-scale merchandising of Raggedy Ann and Andy during the 1930s.

But company partners were determined to change all that. Guided by his late friend's artistic sensibilities and by his own marketing instincts and keen sense of the Raggedys' traditional and commercial appeal, Cox embraced the role of managing partner with great enthusiasm and emerged as the driving force behind the Johnny Gruelle Company.

The first order of business was to issue a new Raggedy Ann and Andy book. Mindful of the necessity of close cooperation with the Donohue Company (and with what he termed "very limited financial resources"), Cox edited and published *Raggedy Ann in the Magic Book*. Based on a manuscript Gruelle had left behind and featuring new illustrations by Johnny's older son, Worth, this new book was in stores in time for Christmas 1939.

When the Johnny Gruelle Company published its first Raggedy Ann book in 1939, this display was mounted in Wanamaker's Department Store window. (Cox-Scheffey Collection)

Official logo of the Johnny Gruelle Company.

THE JOHNNY GRUELLE COMPANY BOOKS

Less than a year after its formation, the Johnny Gruelle Company began issuing new Raggedy Ann and Andy books based on manuscripts that Gruelle had prepared to be used first as serials for newspaper publication, and later, as the basis for new books. The roster of new books would grow to include *Raggedy Ann in the Magic Book* (1939), *Raggedy Ann and the Golden Butterfly* (1940), *Raggedy Ann and Andy and the Nice Fat Policeman* (1941), *Raggedy Ann and Betsy Bonnet String* (1943), and *Raggedy Ann in the Snow White Castle* (1946).

(161)

(162)

(163)

The Johnny Gruelle Company reprinted most of Gruelle s original Volland Happy Children titles.

The Gruelle Company would eventually reprint Johnny s original Volland titles as well all except *Raggedy Andy s Number Book, The Magical Land of Noom, Wooden Willie, Raggedy Ann s Magical Wishes, Marcella, Raggedy Ann s Lucky Pennies*, and his six Sunny Book titles.

While resembling previous Volland and Donohue editions, the slightly wider Johnny Gruelle Company books possess their own distinctive hallmarks, including solid color, rather than illustrated, endpapers incorporating the official Johnny Gruelle Company logo (the Raggedys sitting inside a circle). Pages are printed on high-gloss coated paper, and editions and printings are unstated.

The Gruelle Company books are illustrated on front covers only; the solid-color back covers feature the Gruelle Company logo. Between 1939 and 1950, the Gruelle Company books came with visible cloth spines in solid colors (the five new titles each having a different color spine and color-coordinated endpapers; the reprint titles with cloth spines in red or black). Between 1951 and 1960, editions began appearing with paper-covered spines on which the book s title was printed. The Gruelle Company books were issued in illustrated dust jackets that duplicated cover artwork.

The Johnny Gruelle Company regularly advertised its books in trade publications such as Publishers Trade List Annual. *(Joel Cadbury)*

(164)

(165)

(166)

(167)

(168)

The five new titles published by the Johnny Gruelle Company.

Next on the agenda was character licensing. After releasing agent Fred Wish (whom Cox felt had not succeeded in furthering the reputation and financial potential of the Raggedys), he enlisted William C. Erskine to handle licensing agreements. Erskine, who had worked for Walt Disney (and would eventually handle such popular characters as Uncle Wiggily, Nancy Drew, and Little Lulu), was an expert in copyright merchandising. The Gruelle Company also recruited Johnny s 27-year-old son, Worth, and 50-year-old brother, Justin, whose skillful artistry would grace many licensed games, toys, and books of the 1940s.

In 1940, partners of the Johnny Gruelle Company met with Fleischer Studios, assigning them exclusive motion picture rights to the Raggedys. The spring 1941 release of the feature-length *Raggedy Ann and Raggedy Andy* was accompanied by a wealth of tie-in merchandise, including a first-time-ever Camel with the Wrinkled Knees doll produced by Georgene Novelties. Along with the cartoon, these toys, clothing, handkerchiefs, bed linens, games, playsets, dishes, and book adaptations introduced a new generation to the extended Raggedy family.

I am decidedly poor at lettering, however, so, if you can arrange to get someone to do the lettering, I am certain that I can give you some lovely drawing to submit.

Johnny Gruelle to Fred Wish
1935

With its coordinated merchandising and impressive list of licensees (which would soon number more than thirty), the Johnny Gruelle Company made it abundantly clear that it was not simply a juvenile publisher. A full-page article in the March 1941 issue of *Playthings* focused on the success of Raggedy Ann as a licensed image:

Not to be overlooked in the consideration of Raggedy Ann merchandise, stressed *Playthings* editors, is the fact that the present young generation of mothers, constituting the principal part of the toy buying consumer public, were themselves brought up on Raggedy Ann. Thus, the wholesome atmosphere which pervades the Raggedy Ann stories is further extended by parent endorsement of Raggedy Ann products.

CALICO MILLIONAIRES: THE RAGGEDYS IN THE MOVIES

In view of Raggedy Ann's and Andy's popularity, it is odd indeed that no significant film release starring them resulted during Johnny Gruelle's lifetime. However, it wasn't for lack of trying.

In May 1924, Old Dominion Motion Picture Corporation had secured rights from Gruelle for a Raggedy Ann and Andy film; however, distributors did not find whatever resulted funny enough to invest in. In late 1933, RKO expressed interest in producing an animated Raggedy cartoon, but the production never materialized.

Later on during the 1930s Gruelle and his agent entertained requests to provide artwork for an unspecified motion picture already in production. They communicated with an H. Emerson Yorke about proposed cartoon shorts to be entitled "Raggedy Ann and Andy Grams." They also considered a proposal from Hal Roach to use the Raggedy characters in "Our Gang." None of these projects yielded tangible results either.

It would not be until after Gruelle's death that the movie-going public would meet the Raggedys. Following a turndown by Walt Disney (and encouraged by early sales figures for the just-published *Raggedy Ann in the Magic Book*) on April 18, 1940, the Johnny Gruelle Company granted exclusive motion picture rights to Fleischer Brothers Studios to use, excerpt, adapt, and translate from any or all Raggedy Ann and Andy storybooks and newspaper stories, for the purposes of producing a Raggedy Ann and Andy movie.

The result was an ambitious Technicolor musical cartoon entitled "Raggedy Ann and Raggedy Andy." Despite its considerable charm and innovation, however, the Paramount-released "Raggedy Ann and Raggedy Andy" failed to generate the hoped-for critical acclaim and public interest. Consequently, neither Fleischer nor Paramount moved forward with a planned series of additional two-reel cartoons featuring Gruelle's rag dolls.

By late 1942, Paramount Pictures had taken over the financially troubled Fleischer Brothers Studios, moving the down-sized operation back to New York City and producing cartoons under its new name, Famous Studios. During the next few years, Famous Studios produced two Raggedy cartoon shorts—"Suddenly It's Spring" (1944) and "The Enchanted Square" (1946), both directed by veteran animator Seymour Kneitel, whose credits included the earlier Fleischer Brother's Raggedy film as well as Betty Boop. However, while conveying the lovable essence of Raggedy Ann and Andy, the Famous Studios cartoons were too sentimental and melodramatic for 1940s moviegoers.

(169)

The marketing campaign for Fleischer/Paramount's "Raggedy Ann and Raggedy Andy" included posters, ads, and tie-in Raggedy merchandise.

(170)

(171)

These upbeat, Broadway-style songs composed especially for "Raggedy Ann and Raggedy Andy" captured the playfulness of Gruelle's characters.

(172)

Children's photo album manufactured by Krueger, Inc. (Brenda Milliren)

(173)

Issued in 1941, the Fisher-Price "Push-and-Pull" toy (#711) featured Raggedys that beat a drum when the toy was rolled. (John Murray/Bruce Fox)

THE MCCALL PATTERNS

In July 1940, the Johnny Gruelle Company authorized the 70-year-old McCall Pattern Company to be the exclusive manufacturer of Raggedy Ann and Andy sewing patterns. Standardized home-use-only patterns were a prudent way of controlling and capitalizing on the thousands of Raggedy dolls home seamstresses were making anyway. It was also a well-timed tie-in to the soon-to-be-released Paramount cartoon "Raggedy Ann and Raggedy Andy."

The McCall Raggedy doll patterns were especially popular during and after World War II. In 1951, the McCall Company considered introducing a line of new patterns for the Marcella, Beloved Belindy, and Camel with the Wrinkled Knees characters, but no new pattern resulted. However, McCall would become a long-time licensee, issuing revised and updated patterns for both Raggedy Ann and Andy as well as Raggedy craft projects, clothing, and costumes. The McCall patterns issued under the Johnny Gruelle Company license include:

"Raggedy Ann and Raggedy Andy" (#820)
19" dolls and clothing, including a cape for Raggedy Ann. (1940; 1958) 35 cents

"Asleep and Awake Raggedy Ann and Andy, Camel with the Wrinkled Knees" (#914)
13" Raggedy Ann and Andy dolls and clothing; 12" camel with blanket. (1941) 35 cents

(176)

(177)

McCall Pattern #820 was issued steadily through the 1950s. Because of Georgene Novelties' exclusive license for commercial Raggedy dolls, McCall stipulated that its patterns were for individual use only, not manufacturing purposes.

(174)

Homemade Raggedys (right) from the 1920s and 1930s were made without patterns. When McCall introduced its Raggedy patterns, homemade dolls (bottom) became more standardized, though seamstresses never hesitated to use their imaginations regarding materials.

(175)

(178)

(179)

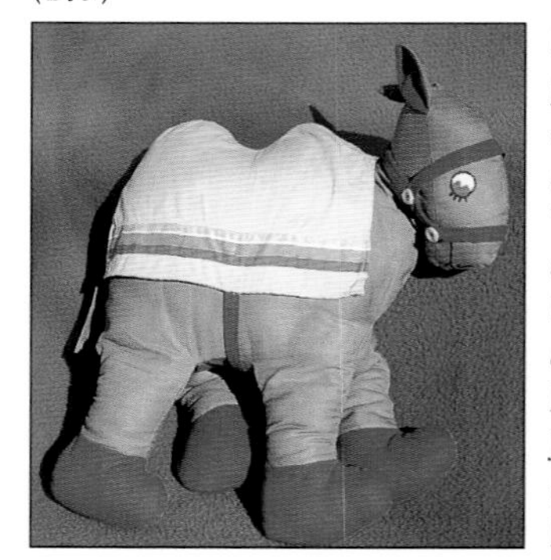

In 1941, McCall issued pattern #914 for Asleep and Awake Raggedys and the Camel with the Wrinkled Knees. These designs closely resembled the dolls being produced by Georgene Novelties, as well as Justin Gruelle's illustrations for the 1940s Raggedy books.

"Raggedy Ann Sunsuit and Panties" (#966)
Toddler's two-piece outfit in sizes 2-6. (1942) 30 cents

"Raggedy Ann and Andy Quilt" (#1063)
Quilt pattern with transfers. (1943)

"Raggedy Ann and Andy Designs for Dish Towels" (#1287)
Iron-on kaumograph transfers for embroidering day-of-the-week towels. (1946) 25 cents

During the early 1940s, McCall produced pattern #1063 for a Raggedy Ann and Andy embroidered quilt.

(180)

McCall transfer pattern #1287, "Raggedy Ann and Andy Designs for Dish Towels," was first issued in 1946.

McCall Style News *(1941). (Marge Meisinger)*

Despite the undeniable success of the Raggedys, by September 1941 partners of the Johnny Gruelle Company had fallen into deep disagreement over how the company should be run to maximize its potential. A contract addendum was drawn up, stipulating a repayment schedule and calling for a cooling-off period before partners decided the ultimate fate of the company.

Then, in December, the United States found itself at war once again. Though materials such as metal, rubber, and wool grew to be in especially short supply, companies specializing in ceramics, printed wooden toys, lithographed products (such as greeting cards, games, paper dolls, coloring books), and cotton textiles (including cloth dolls, clothing, costumes, and linens) continued their production of toys and character merchandise. Consequently, the period coinciding with World War II was one of the richest in Raggedy merchandising history.

However, time had not solved the internal problems plaguing the Johnny Gruelle Company. The partners were forced to admit that a new arrangement was needed and that incorporation was a necessity if the company hoped to maintain a strong presence in the playthings marketplace.

> ***"And our cheery smiles are painted on, too," Raggedy Ann added. "We even smile in our sleep."***
> ***Raggedy Ann and the Left Handed Safety Pin***
> **1935**

In October 1943, the partnership was dissolved. Managing partner Howard Cox made a cash investment and became president of the newly incorporated Johnny Gruelle Company. Under a license from Myrtle Gruelle, Cox would now oversee all publishing and sub-licensing of the Gruelle copyrighted and trademarked material (except for the stuffed Raggedy Ann and Andy dolls, which Myrtle Gruelle would continue administering directly). Though set royalty percentages would go to the Gruelle family, who also retained final approval on licensees, the Johnny Gruelle Company was now in the hands of Howard Cox.

In December 1944, Myrtle Gruelle further distanced herself from the business of the Raggedys,

THE GEORGENE CAMEL WITH THE WRINKLED KNEES

Georgene Novelties began producing its Camel with the Wrinkled Knees doll as a planned tie-in to the newly released Paramount-Fleischer cartoon "Raggedy Ann and Raggedy Andy," which featured Gruelle's daffy book character. Licensed on March 18, 1941, the Georgene Camel was the first commercial doll to be patterned after Gruelle's Camel. The Volland Company's earlier disinterest may have stemmed from Gruelle's admission that he had appropriated his camel idea from an already existing commercial toy.

The affable-looking Georgene Camel stands approximately 10" tall on his four legs. He is made from felted tan cloth, with red fabric feet, a red bridle with four tin buttons, and yellow blanket set off with red and turquoise fabric stripes and tin buttons. His "googly" eyes are celluloid (other features were printed on) and he possesses no company markings. Like the Georgene Raggedy Ann and Andy and Beloved Belindy dolls of that era, the Camel with the Wrinkled Knees dolls retailed for around $2.

Though its Camel with the Wrinkled Knees was likely eased out of production much earlier, Georgene Novelties' license for the doll was renewed annually until 1954.

(182)

(181)

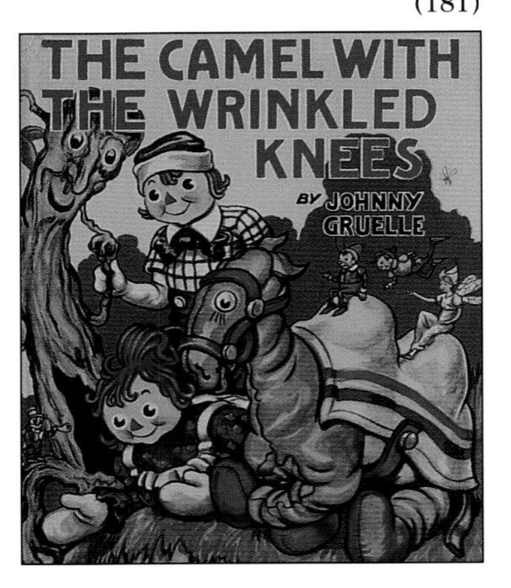

The introduction of the Georgene Camel with the Wrinkled Knees (top) in summer 1941 was timed to coincide with the newly released "Raggedy Ann and Raggedy Andy" cartoon (bottom) and the August publication of McLoughlin Bros.' The Camel with the Wrinkled Knees *(left). (Pauline Baker/Cleveland State University Library/Author's Collection)*

We're on our way to drink pink tea in Granny Hootie Owl's picnic tree!

Raggedy Ann + Andy
Boys' and Girls' March of Comics
1948

designating attorney William G. Fennell as her personal agent with power of attorney to grant or withhold licensing decisions on her behalf. It was also around this time that the artwork of Worth and Justin Gruelle was replaced with that of other, non-family artists for the books and licensed merchandise featuring the Raggedys.

(183)

"Raggedy Ann and Raggedy Andy Sticker Kit Circus" (#546) by Kits, Inc., contained tracing patterns, colored gummed paper, and mounting cardboard, along with a tracing stylus, rubber sponge, and other supplies to make a toy circus.

Howard Cox never deluded himself about what he brought to the company named after his late friend. Cox was neither an artist nor a writer and had never purported to be. And he never considered himself a replacement for Johnny Gruelle. But he knew he possessed the know-how, stamina, and connections to make Raggedy Ann and Andy a going business, one that could earn income for the Gruelles and himself.

Cox spent the remainder of the 1940s overseeing book publishing and reprinting, as well as tending to company finances. William Erskine continued

---

"What fun!" cried Andy. "And no girls around to tell us to be careful!"

Raggedy Andy Goes Sailing
1943

THE GEORGENE BELOVED BELINDY

Eager to introduce more Gruelle characters as dolls, on November 1, 1938, the Johnny Gruelle Company authorized Georgene Novelties to begin manufacturing a newly designed Beloved Belindy doll. Though it is not clear exactly when these dolls reached the marketplace, by the early 1940s the Georgene Beloved Belindy doll was widely available.

The doll functioned as a tie-in to the Donohue edition of Gruelle's *Beloved Belindy,* which the Johnny Gruelle Company would not take over reprinting until 1952. The Georgene license to produce Beloved Belindy dolls expired during the mid-1950s; however, the Johnny Gruelle Company was still recording royalty income for the doll as late as July 1960.

The Georgene Beloved Belindys can be grouped into several categories, according to date(s) of manufacture, style, and size.

(Note: Doll identifiers that appear in quotation marks are either commonly used terms or the author's nomenclature—not official company designations. They are presented here to assist in identifying and classifying representative Georgene Beloved Belindy dolls.

"Standard" Beloved Belindys (ca. 1939-late 1940s)

Made of medium-brown fabric, the first Georgene Beloved Belindys (referred to here as the "Standard" doll) stands 18" tall and possesses a small waist, plump tummy, wide hips, and a round, flat head. Every bit as embraceable as the earlier Volland version, the Georgene "Standard" Belindy has silk-screened facial features that include a white mouth and eyebrows and a red nose, all outlined in black. Her widely set eyes are large, white shirt buttons.

The "Standard" Beloved Belindy's red-and-white-striped leggings go up only as high as her unjointed knees, and her cloth feet (which turn out to the sides) are made of solid red fabric. She is dressed in removable clothing that includes a yellow-and-red floral-print skirt, red blouse with white eyelet collar and four yellow metal buttons, white pantalettes, and white apron. Like her Volland predecessor, she wears a bandanna in place of hair, which is held on with several small safety pins.

An identifying stamp (appearing either on the back of her head or on her bottom) reads: "Johnny Gruelle's Own 'Beloved Belindy.' Trademark Reg U.S. Patent Office. Copyright 1926 John B. Gruelle. Georgene Novelties, Inc. NYC. Exclusive licensed Manufacturer." She also came with a heart-shaped paper hang tag.

During World War II, "Standard" Beloved Belindys appeared with variations in fabric design, including polka-dotted feet, bandannas, and blouses.

"Little" Beloved Belindys (mid-1940s-late 1940s)

During the mid-1940s, Georgene introduced a 15" doll (referred to here as "Little" Beloved Belindy). These scaled-down dolls echo the detailing of the 18" "Standard" dolls, including bottoms that were specially constructed for sitting, and the fabrics used for body and clothing were markedly lighter and more muted than the larger dolls.

"Forward Feet" Beloved Belindy (late 1940s-50s)

By the late 1940s, a restyled Beloved Belindy made of rich brown fabric was offered. This less voluptuous, more streamlined doll (which, like the earlier "Little" Belindy, stands 15") wears a blouse with two center metal snaps (instead of four buttons), has no black nose outline, and lacks company body markings. Her most prominent feature (hence her "Forward Feet" designation) is her boxy cloth feet which, instead of pointing to each side, face front.

(184)

(185)

Georgene Novelties manufactured Beloved Belindys in several sizes and styles. Shown here are (left) the 18" "Standard" Belindy from the 1940s and (right) the 15" "Forward Feet" Belindy from the late 1940s-early 1950s. (Pauline Baker/Author's Collection)

BELOVED BELINDY

658. That famous colored character in the Raggedy Ann stories. A big, soft rag doll, 19 inches tall. Button eyes, with bright cotton clothes to take off, to wash, to put on. Made with the most painstaking attention to details. Children love to take Belindy to bed with them............$2.25

The "Standard" Georgene Beloved Belindy was advertised in the September 1941 Robert Kellogg catalogue. (Marge Meisinger)

(186)

White & Wykoff produced Raggedy greeting cards like this one (#T 9366) dating from 1941. Among the company's other Raggedy cards were "Happy Birthday—Here We Are All Spic and Span . . ." (#T 9363) and "Raggedy Andy and Raggedy Ann are dashing around as fast as they can" (#T 5040).

handling licenses, which, to ensure uniqueness and quality, were exclusive and granted only to companies whose workmanship and reputations were well known. Licensing agreements stipulated royalties ranging from 1 to 10 percent of either wholesale or retail price (depending on type and price of the product), with most set at a standard 5 percent. Occasionally a license would reflect a simple copyright assignment, a clarification of rights, or a one-time payment. More than once, an infringement was transformed into a paid license. Cox kept detailed records, and the licensed Raggedy merchandise of the 1940s sold well.

Georgene Novelties continued producing Raggedy Ann and Andy dolls throughout the 1940s, and the dolls remained front runners, even while the company was busy marketing other popular character dolls. As might be expected, styles and designs of the Georgene Raggedy dolls changed, but Georgene's attention to quality was steadfast.

continued on page 136

THE MILTON BRADLEY PLAY SETS, PUZZLES, AND GAMES

When the Johnny Gruelle Company granted a license to Milton Bradley in June 1940, it was connecting with a playthings industry kingpin. Milton Bradley had founded his company more than eighty years before in Springfield, Massachusetts, building a reputation on high-quality lithographed games and toys. The company would produce a wide variety of authorized Raggedy merchandise, including picture puzzles, paper dolls, board games, and play sets.

Under its license, Milton Bradley agreed to pay the Johnny Gruelle Company royalties ranging from 5/8-cent to 2 1/2 cents, based on retail prices, and could renew its contract each year provided royalties amounted to at least $1,000 annually. Milton Bradley's license would be among several longtime Gruelle Company contracts that would pass to Bobbs-Merrill in the 1960s.

The following authorized merchandise was among that issued by Milton Bradley:

Raggedy Ann Picture Puzzles (#4855)
Four 10" x 13" jigsaw picture puzzles, lithographed and wrapped in glassine, in lithographed box. Artwork by Worth Gruelle and unidentified artist. Box measures 10 3/8" x 13 3/8". (1940) 75 cents

The Games of Raggedy Ann (#4851)
Children's board game, with lithographed folding board, enameled-wood playing pieces in four different colors, two dice, in lithographed box with build-up and partition. Artwork by Worth Gruelle. Box measures 9 1/2" x 19". (1941) $1.50

Raggedy Ann's Magic Pebble Game (#4865)
Children's board game with lithographed playing board, wood enameled playing pieces, and die, packaged in lithographed box. Artwork by Worth Gruelle. Box measures 8 3/4" x 15 1/2". (1941) 35 cents

Raggedy Ann Cutout Dolls With Dresses (#4106)
Raggedy Ann and "three modern misses" paper dolls lithographed on heavy cardboard, with stands and eight sheets of costumes to cut out. Packaged in lithographed box with build-up. Box top artwork by Justin Gruelle. Raggedy Ann paper-doll design patterned after Johnny Gruelle's *Raggedy Ann Cut-Out Paper Doll* (Whitman, 1935). Box measures 12 1/2" x 14". (1941) $1

Raggedy Ann's Bubble Set (#4867)
Two wooden pipes, two cakes of soap, two green, enameled water pans, and rubber apron, in partitioned lithographed box. Artwork by Worth Gruelle. Box measures 9" x 14". (1941) 75 cents

(188)

(189)

(190)

Milton Bradley Raggedy puzzles (#4855-A) and games (#4851, #4865) featured artwork by Worth and Justin Gruelle. (Brenda Milliren/Author's Collection)

(187)

Milton Bradley's "Raggedy Ann Cutout Dolls" featured Raggedy Ann, Marcella, and "three modern misses."

Raggedy Ann's Crayon Set (#4860)
Twenty-four 3½" x 5½" sheets of paper with Raggedy pictures (three different subjects) to color, twenty-four small and seven large crayons in six standard colors and black, packaged in lithographed box with partitioned sections covered in glassine. Box measures 10½" x 12". (1941) 75 cents

Raggedy Ann's Embroidery Set (#4894)
A large set with six cloth doilies (assorted colors and sizes) stamped with Raggedy Ann characters, scissors, two wood embroidery hoops, six spools embroidery cotton, package of needles, and thimble, packaged in lithographed box with build-up. Artwork by Justin Gruelle. Box measures 13" x 17". (1941) $1.25

Raggedy Ann's Embroidery Set (#4893)
A medium-size set with five doilies, four skeins embroidery cotton, wooden embroidery hoops, scissors, thimble, and two needles, packaged in lithographed box with build-up. Artwork by Justin Gruelle. Box measures 10½" x 15¾". (1941) 75 cents

Raggedy Ann's Embroidery Set (#4892)
A beginner's set, with one 9" x 9" doily and four 6" x 6" doilies, wooden embroidery hoops, two skeins embroidery cotton, and needle, packaged in lithographed box with build-up. Artwork by Justin Gruelle. Box measures 9½" x 13¼". (1941) 35 cents

Raggedy Ann and Andy Paint and Crayon Set (#4905)
A jumbo set with six large pictures (9" x 12") and assorted small pictures (3½" x 4½") to color and paint, nineteen large and eight small wood cups of paint, two tubes watercolors, eight wax crayons, two white, enameled water pans, brush, and color mixing chart. Packaged in lithographed box with build-up and hinged lid. Artwork by Justin Gruelle. Box measures 13" x 18". (1941) $1.50

Raggedy Ann and Andy Paint and Crayon Set (#4904)
A medium-size set with twelve pictures (3¾" x 5") to color, twenty-three hard-cake watercolors, eight assorted crayons, two white, enameled water pans, brush, and color mixing chart. Packaged in box with build-up and lithographed lid. Artwork by Justin Gruelle. Box measures 11½" x 16¼". (1941) $1

Raggedy Ann and Andy Paint and Crayon Set (#4903)
A small set with six Raggedy Ann outline pictures (3½" x 5¼"), fourteen hard-cake watercolors, five wax crayons, two blue, enameled water pans, brush, and color mixing chart. Packaged in box with build-up and lithographed lid. Artwork by Justin Gruelle. Box measures 9¾" x 15". (1941) 50 cents

continued on next page

10 MILTON BRADLEY GAMES

RAGGEDY ANN

Games
Puzzles
Painting
Crayoning and
Embroidery Sets

4851 RAGGEDY ANN GAME

4865 RAGGEDY ANN'S MAGIC PEBBLE GAME

4867 RAGGEDY ANN BUBBLE SET

4855 RAGGEDY ANN PICTURE PUZZ

4860 RAGGEDY ANN'S CRAYON SET

MILTON BRADLEY GAMES 11

RAGGEDY ANN

4894 RAGGEDY ANN'S EMBROIDERY SET

4893 RAGGEDY ANN'S EMBROIDERY SET

4892 RAGGEDY ANN'S EMBROIDERY SET

4905 RAGGEDY ANN AND ANDY PAINT AND CRAYON SET

4904 RAGGEDY ANN AND ANDY PAINT AND CRAYON SET

4903 RAGGEDY ANN AND ANDY PAINT AND CRAYON SET

Milton Bradley advertised an entire line of licensed Raggedy Ann and Andy merchandise in its 1940s sales catalogues.

MILTON BRADLEY GAMES 29

4472—RAGGEDY ANN AND ANDY RAILROAD PUZZLE

The beloved Raggedy Ann and her constant companion, Andy, of cartoon fame are portrayed on their long train, with associated figures of this popular rag doll character. Lithographed in striking colors and mounted on heavy board. Comes assembled in six sections, each measuring 6¾ x 18 inches, with different colored back paper. When put together the train measures 108 inches long. Makes a very attractive counter display piece. Lithographed label.

Size 13 x 18¼. (6) Price, each, $1.50

4855-A—RAGGEDY ANN PICTURE PUZZLES

The delightful Raggedy Ann characters in action, lithographed and mounted on heavy cardboard, scroll cut into many pieces. Three different pictures, each 10 x 13 when assembled, and each with its individually colored back for easy identification of pieces. Assembled in box. Lithographed label.

Size 10⅝ x 13⅞. (24) Price, each, $0.60

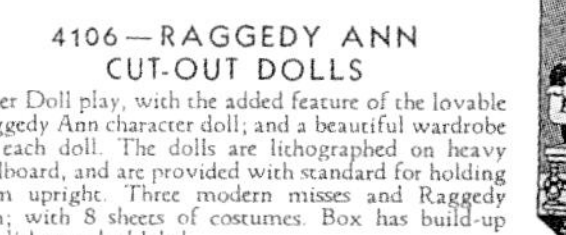
4106—RAGGEDY ANN CUT-OUT DOLLS

Paper Doll play, with the added feature of the lovable Raggedy Ann character doll; and a beautiful wardrobe for each doll. The dolls are lithographed on heavy cardboard, and are provided with standard for holding them upright. Three modern misses and Raggedy Ann; with 8 sheets of costumes. Box has build-up and lithographed label.

Size 12½ x 14. (12) Price, each, $1.00

Raggedy Ann Picture Puzzles (#4855-A)
Three different 10" x 13" jigsaw puzzles, with artwork by Worth and Justin Gruelle. Box measures 10 3/8" x 13 3/8" (some sets also issued in slightly smaller boxes measuring 9 3/8" x 12 1/4" with words "Raggedy Ann" added to box side). (1944) 60 cents

Raggedy Ann's Magic Pebble Game (#4865-A)
Children's board game. Same as #4865 except box size reduced to 8 5/8" x 13 1/4". (1944) 50 cents

Raggedy Ann and Andy Railroad Picture Puzzles (#4472)
Six lithographed jigsaw puzzles measuring 6 3/8" x 18", which when placed end-to-end created a 108" Raggedy railroad train. Packaged in lithographed box. Artwork by Justin Gruelle. Box measures 13" x 18 1/4". (1944) $1.50

Raggedy Ann: A Little Folk's Game (#4809)
Children's board game. Full-color lithographed playing board, spinner, cards, and box lid. Lithographed box measures 8 1/2" x 16 1/2". (1954)

Raggedy Ann—3 Interlocking Puzzles (#4500)
Three interlocking 9" x 12" jigsaw puzzles lithographed in color. Lithographed box measures 9 3/8" x 12 1/4". (1954)

Raggedy Ann Puzzle (#4508-4)
Twenty-piece lithographed jigsaw puzzle (image is the Raggedys on a picnic). One of sixteen "aptitude-tested" puzzles for ages 4-9. (1955)

Raggedy Ann and Andy Dolls (#4938)
Punch-out lithographed figures with colorful backgrounds. Lithographed box measures 7" x 10". (1959)

Raggedy Ann—2 Elementary Puzzles
Two interlocking 9" x 12" jigsaw puzzles lithographed in color. Lithographed box measures 9 3/8" x 12 1/4". (1965)

(191)

Milton Bradley introduced this updated Raggedy Ann and Andy game in March 1954.

(192)

Milton Bradley tray puzzle (1955).

THE CROOKSVILLE RAGGEDY ANN AND ANDY WARE

The Crooksville China Company secured its license for Raggedy Ann dishes on December 2, 1941, only days before the U.S. entered World War II. This 39-year-old Ohio-based pottery specialized in vases, flower pots, and other novelty items, as well as china place settings, accessories, and children's dish sets.

The Crooksville "Raggedy Ann and Andy Ware" included a child's mug, child's bowl, compartment plate, and baby plate, which wholesaled for $1.60, $1.80, $2.25, and $2.25, respectively, per dozen. The sturdy glazed pieces are decorated with images of Raggedy Ann, Raggedy Andy, the Camel with the Wrinkled Knees, and Uncle Clem. Edges of all four pieces are outlined with a narrow gold or red stripe.

Though most pieces are stamped on the bottom with "Crooksville Raggedy Ann and Andy Ware by Crooksville. Copyright 1941 Johnny Gruelle Co," some pieces escaped the factory without a stamp. The Johnny Gruelle Company received a 5 percent royalty on sales of the Crooksville "Raggedy Ann and Andy Ware."

Imprint from the Crooksville Raggedy Ann and Andy Ware.

(193)

The Crooksville Raggedy Ann and Andy Ware included a child's mug, child's bowl, baby plate, and compartment plate. The unusual gingham-and-print designs of the Raggedys and Camel on the Crooksville dishes departed from the usual 1940s authorized artwork.

(194)

(195)

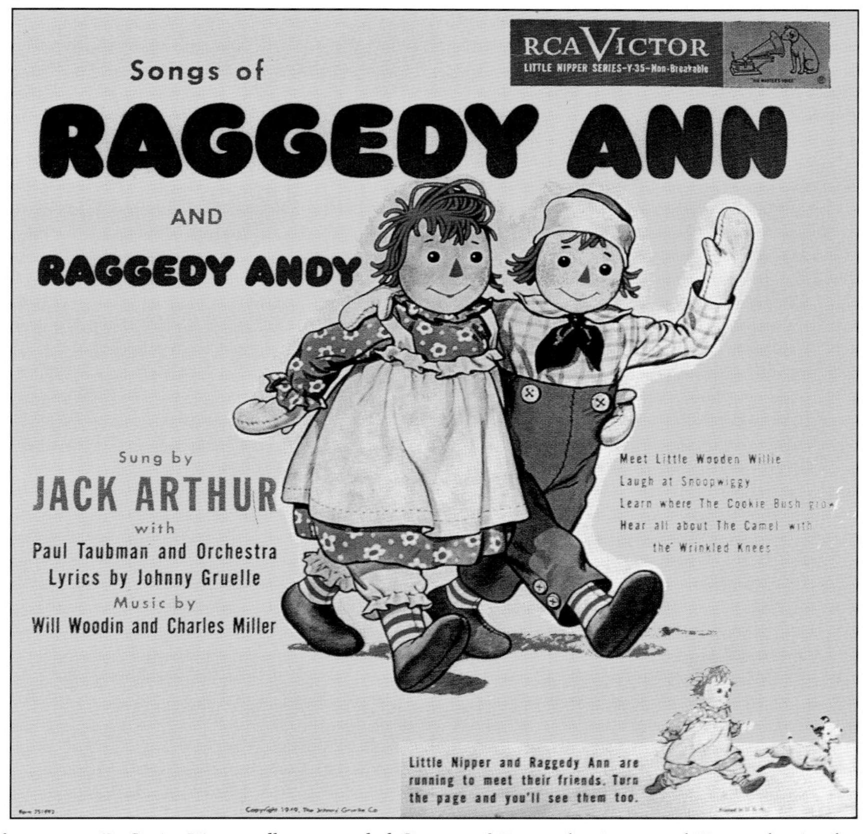

In 1948, Jack Arthur recorded arrangements for a new R.C.A. Victor album entitled Songs of Raggedy Ann and Raggedy Andy, *released in both 78 rpm and 45 rpm formats (left). In 1949, the album folder (right) was redesigned. (Author's Collection/Brenda Milliren)*

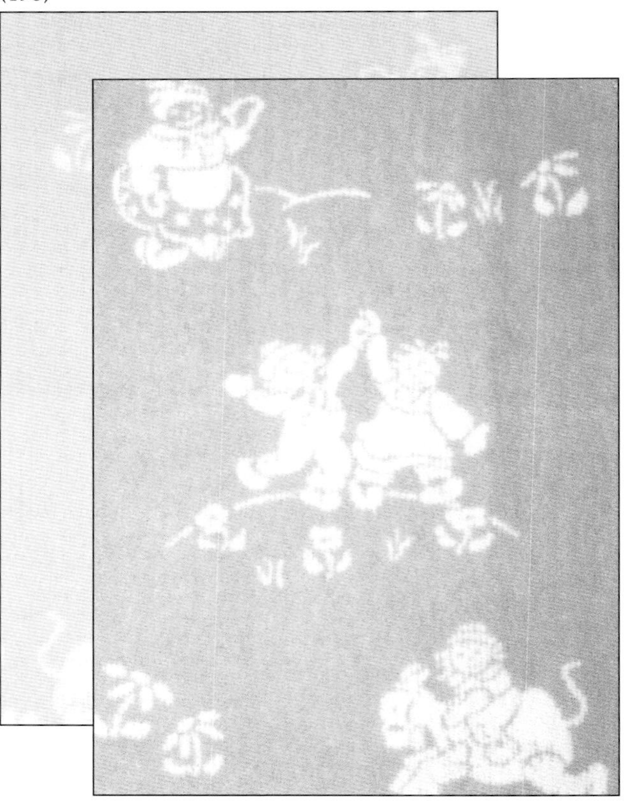

Raggedy doll and baby blankets were produced in the early 1940s by the North-Carolina-based Beacon Manufacturing Company. (Jacki Payne)

In 1945, Barnes Advertising Agency of Milwaukee secured a license to include the Raggedys in billboards advertising root beer. (Cox-Scheffey Collection)

(197)

In 1946, Decca Records issued new arrangements of the Gruelle-Woodin-Miller songs in this three-record set, performed by Frank Luther.

THE HOLGATE TOYS

Holgate Brothers became a Johnny Gruelle Company licensee in April 1941. Founded in 1789 by Cornelius Holgate as a manufacturer of broom handles and store displays, the company by the early 1940s had developed a specialty of wooden educational toys. Key to its superior product line was Holgate's talented designer, Jarvis Rockwell, the brother of Norman.

Holgate manufactured several Raggedy Ann and Andy wooden toys. One, called a "Rocky Toy," is a brightly painted bungalow set atop a roly-poly base depicting Raggedy Ann and Uncle Clem. Another version features Beloved Belindy and Raggedy Andy.

A second Holgate offering, a "Concentration Toy," is a set of two stackable Raggedy Ann and Andy figures that can be taken apart and put back together on their dowels and bases.

A third Holgate Raggedy toy, which may never have been manufactured, was described in a 1941 contract as "a fence made of wooden pickets which, when properly placed and arranged upon their supports, will form a picture of or scene embodying the characters."

In 1946, Holgate opted not to renew its license with the Johnny Gruelle Company.

(198)

Holgate's Rocky Toy was painted wood with a roly-poly base.

(199)

The Concentration Toys (#1204) retailed for $1.50 and came packaged in a box bearing the Holgate name and a commendation from Parent's *magazine. (Suzette Phillips)*

(200)

Children's plastic handbags were produced by Banner Brothers beginning in 1941. In 1948, Georgene Novelties produced a different line of Raggedy toy handbags. (Candy Brainard)

continued from page 129

By the 1950s, however, consumer tastes were shifting to action-oriented, realistic toys. A postwar affluence gave families expanded purchasing power to buy bigger, more elaborate toys and games, many made of sleek, durable plastic. Nostalgia—particularly the kind embodied by Gruelle's Raggedys—was not high on the list of children and parents who saw themselves as technical-minded and forward-thinking.

There was also another strong influence at work during the 1950s molding the tastes and desires of the youngsters who sat raptly watching it for several hours each day: television. Television-inspired games, toys, and dolls came to dominate the playthings marketplace, bearing such images as Howdy Doody, Roy Rogers, and characters featured on a daily television program called "Walt Disney's Mickey Mouse Club."

Although in the late 1940s and early 1950s Raggedy Ann and Andy had been featured on an R.C.A. Victor-sponsored children's radio program,

(201)

(202)

(203)

Raggedy Ann and Andy prints were manufactured during the 1940s by Artograph Company and Georgene Novelties.

(204)

Licensed but unmarked, glazed pottery planters were manufactured by William Hirsch of Los Angeles between 1943 and 1949. Hirsch also produced Raggedy pottery figurines.

His illustrations are painted with an elfin brush dipped in the brilliant colors of the glorious sunshine.
Johnny Gruelle's Golden Book
1941

Gruelle's rag-doll characters never made the transition to television. Licenses secured by Lawrence Wynn and Robert Keeshan (Captain Kangaroo) of Keeshan-Miller during the 1950s never resulted in regular Raggedy Ann television programming. With little or no television exposure and scant on-air advertising for Raggedy merchandise, Gruelle's characters were hard pressed to compete with Hula Hoops, toy cowboy guns, and a doll named Barbie, who made her debut in 1959.

continued on page 147

(205)

(206)

F. A. Foster's 1940s "Puritan" brand drapery fabric featured the Raggedys along with Eugene Field's Gingham Dog and Calico Cat characters. (Cox-Scheffey Collection)

THE MCLOUGHLIN BROTHERS BOOKS, PAPER DOLLS, AND COLORING BOOKLETS

In June 1940, McLoughlin Brothers secured a license from the Johnny Gruelle Company to publish derivative Raggedy Ann and Andy books, coloring booklets, and paper dolls. McLoughlin, one of the oldest American juvenile publishing houses, had been acquired in 1920 by Milton Bradley, which retained not only the McLoughlin imprint for its line of books but also the time-proven services of the company's lithographers and printers.

Though one was the parent company of the other, Milton Bradley and McLoughlin Brothers maintained separate licenses with the Johnny Gruelle Company. While Milton Bradley's license would be long-term, McLoughlin's would be renewed only until 1947, its Raggedy line having ceased several years before that, supplanted by similar merchandise produced by Saalfield.

In addition to Raggedy paper dolls and coloring booklets, McLoughlin issued several series of illustrated Raggedy story books. Issued in 1940 and 1941, the "Little Color Classics" were a half dozen single-episode derivative books based on Johnny Gruelle manuscript material, some previously published as newspaper serials. In November 1943, McLoughlin published reformated editions of all six "Little Color Classics" titles in its "Westfield Classics" series, which sold for 50 cents.

(207)

McLoughlin's Raggedy Ann's Picture Book *was illustrated by Justin Gruelle.*

(208)

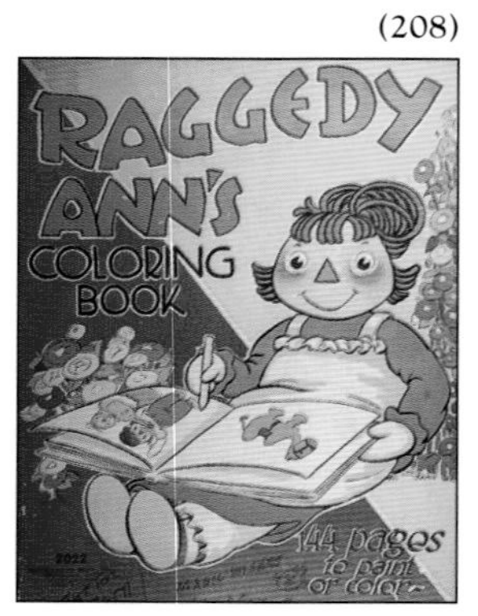

This McLoughlin coloring book featured artwork by Worth Gruelle. In 1941, the Johnny Gruelle Company authorized Hutchinson House of London to publish this book, along with Marcella's Raggedy Ann Doll Book, *for distribution in Great Britain.*

The McLoughlin Brothers Raggedy books and booklets include:

Raggedy Ann's Picture Book (#405)
Limp-cover picture booklet, illustrated by Justin Gruelle. (1940)

Marcella's Raggedy Ann Doll Book (#547)
20-page paper-doll booklet. (1940)

Raggedy Ann's Gift Box (#1020)
Boxed set consisting of "Raggedy Ann's Coloring Book"(#2022), "Marcella's Raggedy Ann's Doll Book"(#547), and a Donohue hard-cover Raggedy Ann book. (1940)

Raggedy Ann's Coloring Book: 144 Pages to Paint or Color (#2022)
Perfect-bound coloring book illustrated by Worth Gruelle. (1940)

"LITTLE COLOR CLASSICS"

Sixty-page derivative books (measuring 6¾" x 5⅛") based on stories by Johnny Gruelle, with black-and-white and color illustrations by Justin C. Gruelle. Titles include:

Raggedy Ann Helps Grandpa Hoppergrass (#843) (1940)
Raggedy Ann and the Hoppy Toad (#844) (1940)
Raggedy Ann in the Garden (#845) (1940)
Raggedy Ann and the Laughing Brook (#846) (1940)
Raggedy Andy Goes Sailing (#883) (1941)
The Camel with the Wrinkled Knees (#884) (1941)

"WESTFIELD CLASSICS"

Forty-four-page editions (measuring 8¾" x 7½") of the "Little Color Classics" titles. Text and illustrations rearranged in a larger, easy-to-read format. Issued in illustrated dust jackets. Titles include:

Raggedy Ann Helps Grandpa Hoppergrass (#60) (1943)
Raggedy Ann and the Hoppy Toad (#61) (1943)
Raggedy Ann in the Garden (#62) (1943)
Raggedy Ann and the Laughing Brook (#63) (1943)
Raggedy Andy Goes Sailing (#64) (1943)
The Camel with the Wrinkled Knees (#65) (1943)

(209) (210) (211)

The 1943 "Westfield Classics" were larger, reformatted versions of all six "Little Color Classics," each issued in a dust jacket.

(212)

(213)

(214)

The first four "Little Color Classics" (bottom), published in 1940, were offered individually and as a boxed set. Two more titles (top) were published in 1941. (Rebecca Greason/photo by Pen Jones)

Very soon they were back in Marcella's nursery, telling the other dolls about their adventure.

Raggedy Ann and Andy—with Animated Illustrations
1944

(215)

(216)

(217)

In 1945, Frank Luther recorded story excerpts from Raggedy Ann and the Golden Butterfly, Raggedy Ann and Andy and the Nice Fat Policeman, *and* Raggedy Ann in the Magic Book. *The records were released in 1952 as singles in both 78 rpm and 45 rpm versions. (Peter Muldavin)*

(218)

(219)

In addition to the train and airport scenes shown here, Braka & Company offered maypole-, merry-go-round-, Halloween-, four seasons-, and "I Love You"-themed Raggedy handkerchiefs during the 1940s. (Cox-Scheffey Collection)

THE LATER VOLLAND GREETING CARDS

Though the Volland Company withdrew from the book and merchandise business in 1934, Gerlach-Barklow retained the Volland imprint for a line of specialty greeting cards. That same year, several Volland Raggedy Ann and Andy cards were introduced but quickly confiscated when Johnny Gruelle (who had not authorized the cards) discovered them and protested.

In November 1940, a bonafide license was secured in Volland's name from the Johnny Gruelle Company to develop eight new children's birthday cards featuring the Raggedys. Several years later, this line was enlarged to include cards for other holidays. Most are dated and numbered on the back and marked "Volland, Joliet, USA." Though the license expired in February 1949, the Raggedy greeting cards were still selling in the early 1950s.

Among these later Volland greeting cards were the following:

"Two Years Old" (#10 F 721) (1940)

"Three Years Old" (#10 F 731) (1940)

"Please Hurry and Get Well" (#10 S 180) (1944)

"Sorry You're Sick" (#10 S 181) (1944)

"It's Your Birthday" (#10 F 887) (1944)

"Happy Birthday to You!" (#10 F 888) (1944)

"A Valentine for a Little Niece" (#10 V 894) (1949)

"Happy Easter to You" (#10 E 178) (1949)

(220)

(221)

Though charming, the artwork on these 5- and 10-cent Volland cards from the 1940s was much more contemporary than the earlier Gruelle-designed valentines. *(Nan Czyzewicz/Author's Collection)*

(222)

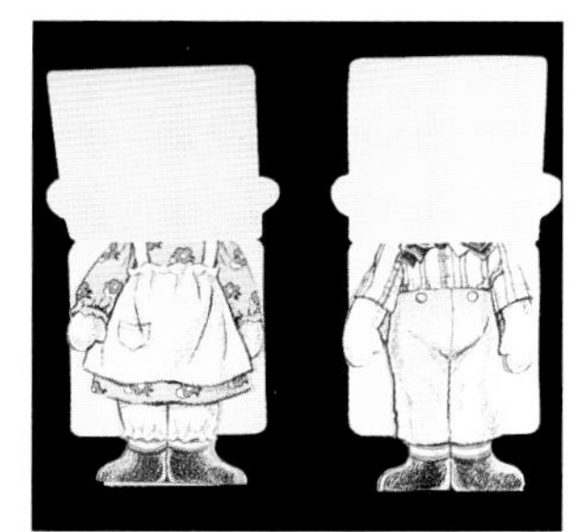

(223)

The designs of these unmarked holiday cards (two with flip-up bottoms) were based on Gruelle's earlier valentine artwork and may have been among the Volland cards issued without his authorization. (Nan Czyzewicz)

The Deep, Deep Woods is a Wonderful Place.
Raggedy Ann and Betsy Bonnet String
1942

THE SAALFIELD BOOKS, PAPER DOLLS, AND COLORING BOOKS

The Saalfield Publishing Company of Akron, Ohio, secured its license with the Johnny Gruelle Company on January 1, 1944. Founded by Albert G. Saalfield, the 44-year-old company was well known for its children's books, paper dolls, and coloring booklets featuring film stars and popular book and comic characters.

In August 1944, Saalfield published *Raggedy Ann and Andy—with Animated Illustrations,* a board-cover storybook starring Raggedy Ann and Andy, the Tired Old Horse, and the Camel with the Wrinkled Knees. This book's unattributed text was loosely based on Gruelle's *The Paper Dragon,* and its clever tab-driven movables had been designed by renowned animator Julian Wehr.

Saalfield's Raggedy line soon grew to include paper-doll and coloring booklets featuring not only both Raggedys but also Beloved Belindy, the Camel with the Wrinkled Knees, Cleety the Clown, Eddie Elf, Snoopwiggy, and even the Quacky Doodles.

The Johnny Gruelle Company reserved the right to approve all Saalfield artwork, most of which fell to veteran artist Ethel Hays <Simms>, whose exuberant, curvilinear style perfectly captured the whimsy and energy of Gruelle's characters. Another artist, Henry Muheim, contributed illustrations for selected Saalfield Raggedy coloring books, patterning his line drawings after those by Gruelle's brother Justin.

(224)

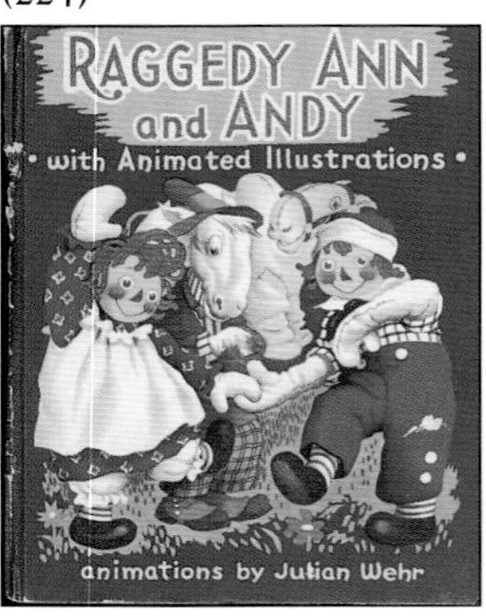

Raggedy Ann and Andy—With Animated Illustrations *(1944) was licensed to Duenewald but published and distributed by Saalfield.*

(225)

Raggedy Ann Picture Story Book *(1947). After reviewing Ethel Hay's cherubic interpretations of the Raggedys, the Johnny Gruelle Company assured Saalfield officials that "your artist has, quite markedly, gotten the Gruelle idea."*

(226)

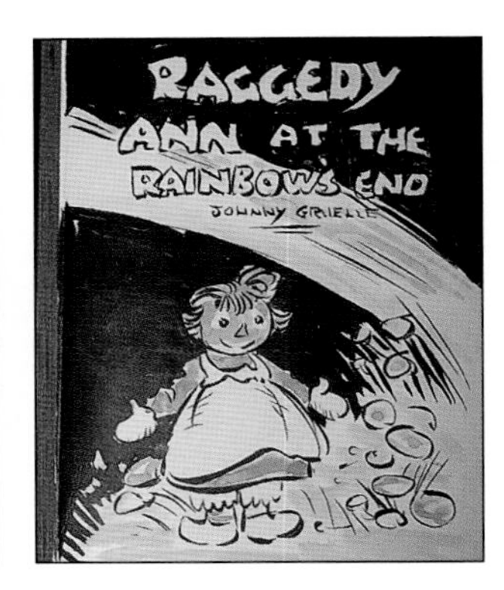

(227)

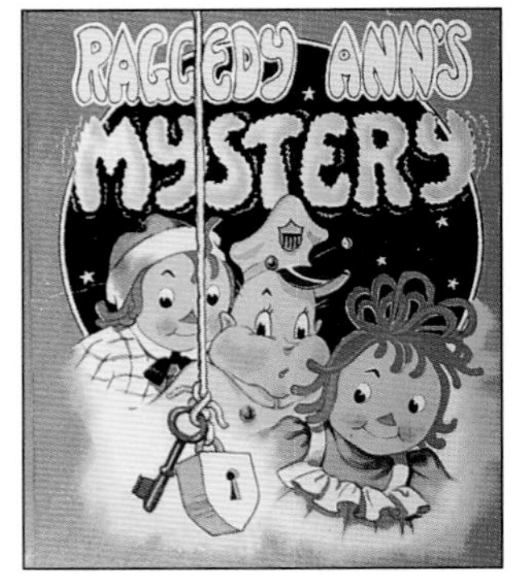

The Saalfield Treasure Books, published nearly a decade following Gruelle's death, were based on Gruelle's unpublished prototype (directly above) created during the 1930s for a mass-market book series. (Author's Collection/Cox-Scheffey collection)

(228)

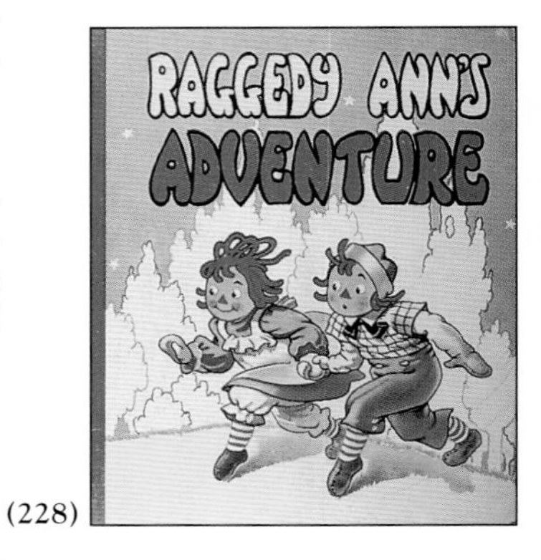

(229)

RAGGEDY ANN AND THE SLIPPERY SLIDE

As was its habit, Saalfield issued and reissued the same Raggedy Ann and Andy booklets over and over, assigning different numbers, rearranging pages, recombining materials and illustrations, and often adding updated covers. While giving the appearance of new publications, this tactic also afforded the company optimum mileage from several sets of artwork.

In 1947, Saalfield published the cloth-like *Raggedy Ann Picture-Story Book,* as well as four derivative Raggedy Ann and Andy storybooks in its "Treasure Book" series. Saalfield continued reissuing its Raggedy

Ann and Andy material until the early 1960s, its license having been taken over by Bobbs-Merrill.

Selected books and booklets published by Saalfield under its Johnny Gruelle Company license include:

Raggedy Ann and Andy—With Animated Illustrations (#858)
Board-cover book with plastic spiral binding and movable illustrations. Animated by Julian Wehr. Issued in dust jacket. (1944)

Raggedy Ann and Andy Paper Dolls (#2497)
Artwork by Ethel Hays. (1944)
Reprints dated 1944: (#369) (#2615) (#2741) (#1506) (#451)
Reprint dated 1950: (#2719)
Reprints dated 1957 (signed Ethel Simms): (#2654) (#2754)

Raggedy Ann Coloring Book (#2498)
Artwork by Ethel Hays. (1944)
Reprint dated 1954: (#1335)

Raggedy Ann and Andy Coloring Book (#370)
Artwork by Henry Muheim. (1944)

Raggedy Ann and Raggedy Andy—Follow the Dots and Color the Pictures (#403)
Four limp-cover coloring booklets in box. Artwork by Henry Muheim. (1945) Titles include:
Raggedy Ann and Raggedy Andy
Raggedy Ann
Raggedy Andy
Cleety the Clown

continued on next page

(230)

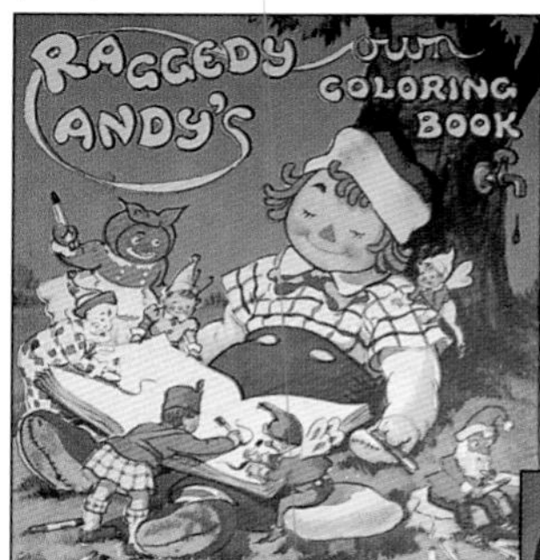

In 1945, Saalfield introduced boxed sets of four coloring booklets, measuring 7¾" x 6½", illustrated by Henry Muheim. (Author's Collection/Brenda Milliren)

(231)

(232)

(233)

(234)

Saalfield's 28-page Raggedy Ann and Andy Coloring Book *(#370) featured the artwork of Saalfield illustrator Henry Muheim, who patterned his drawings after Justin Gruelle's.*

Coloring Fun with Raggedy Ann and Raggedy Andy (#405)

Four limp-cover coloring booklets in box. Box and booklet cover artwork by Ethel Hays. Interior artwork by Henry Muheim. (1945) Titles include:

Coloring Fun with Raggedy Ann and Raggedy Andy
Raggedy Ann's Own Coloring Book
Raggedy Andy's Own Coloring Book
Cleety the Clown's Coloring Book

Raggedy Ann and Raggedy Andy—4 Books to Color (#502)

Four limp-cover coloring booklets in box. Box and booklet cover artwork by Ethel Hays. (1945) Interior artwork by Henry Muheim. (1945) Titles include:

Raggedy Ann's Coloring Book
Raggedy Ann with her Friends—A Book to Color
Raggedy Ann and Beloved Belindy
Raggedy Ann and Raggedy Andy—A Book to Color

Raggedy Ann Picture-Story Book (#2543) (#2929)

Limp-cover book of illustrated verses. Artwork by Ethel Hays. (1947)

(235)

This 32-page Saalfield coloring book (#1310) was published in January 1951.

SAALFIELD TREASURE BOOKS (#463)

Storybooks adapted from Johnny Gruelle's newspaper serials of the 1920s. Artwork by Ethel Hays. Bound in illustrated cardboard covers, with plastic spirals. (1947) Titles include:

Raggedy Ann's Adventure (#463)
Raggedy Ann's Mystery (#463)
Raggedy Ann and the Slippery Slide (#463)
Raggedy Ann at the End of the Rainbow (#463)

Raggedy Ann Coloring Book (#1310)

A "shaped"coloring book. Interior artwork by Henry Muheim. (1951)

(236)

Raggedy Ann and Andy Paper Dolls *(#2741) was among the many reprints of Saalfield #2497, which ranked among the company's top twelve most-reprinted paper doll booklets.*

(237)

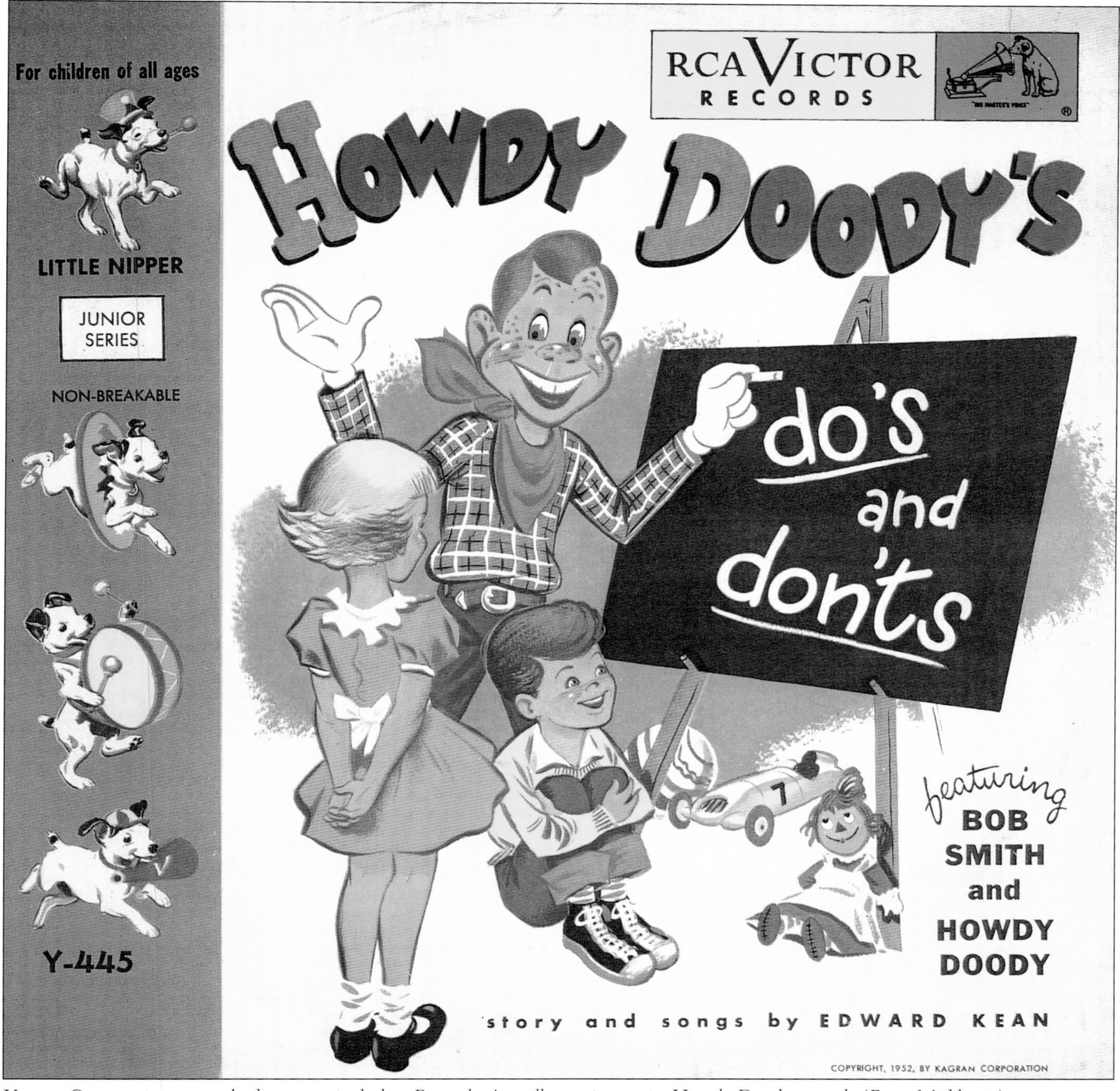

Kagran Corporation secured a license to include a Raggedy Ann illustration on its Howdy Doody record. (Peter Muldavin)

continued from page 138

This market reality led to the downsizing and eventual dissolution of the Johnny Gruelle Company. Throughout the 1950s, the company maintained its New York City address (actually William Erskine's Rockefeller Center office), but Howard Cox had moved the publishing portion of the business to his farm in Princeton Junction, New Jersey.

In early 1959, following a bitter dispute over royalties, Howard Cox and William Erskine ended their eighteen-year association. Though they remained contractually bound to each other, Cox moved all remaining Gruelle Company material out of the New York office, and the two men barely spoke again.

THE RAGGEDY WAY: THE DELL COMIC BOOKS

Though comic drawing had been Johnny Gruelle's forte, Raggedy Ann and Andy never appeared in comic-book form during Gruelle's lifetime. Because comic books, in their paper-cover format, were not introduced until 1933, they may simply have been a medium too new and peripheral for Gruelle.

However, by the early 1940s, modern comic books had come into their own, proliferating in what would later be dubbed their "Golden Age." On March 2, 1942, Dell Publishing secured its first license from the Johnny Gruelle Company, and for the next several decades the Raggedys, along with many of their friends, starred in a variety of Dell comic books.

Comic books certainly fit with the Gruelle Company's ambitious licensing plans of the mid-1940s, and at several points—notably spring 1943 and spring 1945—Gruelle Company material simultaneously appeared in three Dell comic-book series: *Animal Comics*, *New Funnies*, and *Raggedy Ann and <+> Andy*.

The comic features and comic books produced by Dell under its license with the Gruelle Company are:

"Raggedy Ann and Andy"

"Johnny Gruelle's Raggedy Ann"
A monthly feature written by Gaylord Dubois. Artwork by L. Bing and George Kerr.
In *New Funnies*, volume 1, #65 - #111.
(July 1942-May 1946)

"Johnny Gruelle's Mr. Twee Deedle"
A monthly feature written by Gaylord Dubois. Artwork by Justin Gruelle.
In *New Funnies*, volume 1, #69 - #82.
(November 1942-December 1943)

"Merry Meadows"
A bimonthly feature by Gaylord Dubois. Artwork by Justin Gruelle.
In *Animal Comics*, #2 - #6.
(February/March-November/December 1943)

(238)

New Funnies no. 74 (April 1943). Most of the Dell Raggedy Ann and Andy comics were illustrated by veteran comic artist George Kerr, who portrayed Raggedy Ann as a redhead and Andy as a brunette.

Justin Gruelle's "Merry Meadows" premiered in Animal Comics no. 2, (February-March 1943).

In 1943, Justin Gruelle reintroduced his older brother's Mr. "Twee Deedle" to a new generation of children in New Funnies.

(239)

(240)

Raggedy Ann *Four Color Comics nos. 23 and 45 (1943). (Cox-Scheffey Collection/ Author's Collection)*

The pantheon of Gruelle characters appearing in Dell's comic books was inspired by Johnny Gruelle's original Raggedy Ann books and newspaper serials.

Boys' and Girls' March of Comics Featuring Raggedy Ann + Andy *(#23) was 26 pages (including inside covers) and consisted of three Raggedy stories. (Judd Lawson)*

"Raggedy Animals"

A bimonthly feature by Gaylord Dubois. Artwork by George Kerr.

In *Animal Comics*, #10 - #15.

(August/September 1944-June/July 1945)

Four Color Comics

Entire comic books devoted to the Raggedys, issued on an ad hoc basis. Written by Gaylord DuBois and illustrated by George Kerr. Issues include:

Raggedy Ann (#5) (1942)
Raggedy Ann and Andy (#23) (1943)
Raggedy Ann (#45) (1944)
Raggedy Ann (#72) (1945)
Raggedy Ann and Andy (#262) (1950)
Raggedy Ann (#306) (1950)
Raggedy Ann + Andy (#354) (1951)
Raggedy Ann + Andy (#380) (1952)
Raggedy Ann + Andy (#452) (1953)
Raggedy Ann + Andy (#533) (1954)

Raggedy Ann + <and> Raggedy Andy

Regular comic monthlies that included the lead feature "Raggedy Ann," an illustrated poem entitled "The Raggedy Way," a Raggedy "Color Up Picture" or "Fun Page," as well as other non-Raggedy features. Written by Gaylord Dubois. Except for several issues in 1949, artwork by L. Bing and George Kerr.

Volume 1, #1 - #39.

(June 1946-August 1949)

Raggedy Ann + Andy

A 26-page comic book issued by K. K. Publications (a subsidiary of Western Publishing) to department and shoe stores to distribute as a giveaway. Artwork by George Kerr.

"Boys' and Girls' March of Comics" #23. (1948)

Raggedy Ann and Andy

A 100-page comic book. Artwork by George Kerr.

"Dell Giant Comics" #1. (February 1955)

Raggedy Ann and the Camel with the Wrinkled Knees

A 32-page comic book. Artist unknown.

"Junior Treasury" #8. (April 1957)

(241)

Raggedy Ann + Andy *no. 1 (June 1946).*

(242)

Johnny Gruelle's Raggedy Ann and Andy and the Camel with the Wrinkled Knees *was adapted in 1957 in Dell's "Junior Treasury" series.*

(243)

The slotted, punch-out "Notch-Ems" toy, copyrighted 1955 by Grosset & Dunlap, featured numerous Gruelle characters. (Cox-Scheffey Collection)

(244)

This unmarked, see-through ball may have been manufactured by U.S. Fiber & Plastics Corporation, which held a license between 1948 and 1950. (Candy Brainard)

With a licensee list that a decade before had numbered several dozen but now stood at under fifteen, the time seemed right to turn things over to an entity that promised to rekindle interest in the Raggedys. By summer of 1960, Myrtle Gruelle, her sons Worth and Richard, and Howard Cox were signing contracts with the Bobbs-Merrill Company of Indianapolis. The detailed paperwork stipulated that Bobbs-Merrill, or its agent, in addition to handling all Raggedy literary material would oversee licensing for all new merchandise other than dolls.

The Gruelle family would receive royalties based on sales, and Howard Cox and William Erskine would be entitled to a percentage of royalty income yielded by existing licenses. Myrtle designated attorney Herbert J. Jacobi to act as her agent with Bobbs-Merrill to grant or deny new licenses as needed.

(245)

(246)

Famous for wooden dominos and alphabet blocks, Halsam produced Raggedy Ann Safety Blocks in $1\frac{1}{4}$" and $1\frac{5}{8}$" sizes, each stamped with letters, words, a fairy-tale character, and raised images of different Raggedy characters. (Crain Collection/Candy Brainard)

(247)

(248)

The Magic Mirror Movie Company sold its "Red Raven" records along with a mirrored phonograph that made images on the record labels appear to dance. (Cox-Scheffey Collection/Author's Collection)

THE RAGGEDYS IN ADVERTISING

The popularity of Gruelle's books and dolls in the 1920s and 1930s and the ambitious licensing of his characters during the 1940s insured a prominent place for the Raggedys in advertising. In addition to the many space ads directly marketing Gruelle's books, dolls, and licensed novelties or offering them as premiums (in such publications as *Child Life, Pictorial Review, Woman's World* and *Playthings*) were the even more plentiful print ads in which Raggedy Ann and Andy and other Gruelle creations would appear only incidentally.

Gruelle had always been interested in advertising, particularly during the 1930s. He was even willing to depart from his popular rag dolls, suggesting to his agent, Fred Wish:

> There may be possibilities in the advertising field for me to do some things similar to the "Yapps Crossing" cartoons, which I did for *Judge* over a period of ten or twelve years, and the "Yahoo Center" cartoons, which I did for *Life* . . . incorporating the advertised product in the central theme of the picture and using names of prominent retailers on the buildings.

Wish dutifully fielded requests from advertisers (among them, Pepsodent Toothpaste and a brewery that wanted Gruelle's distinctive little gnomes to incorporate in its beer advertising). At one point, Maxwell House Coffee heir William Cheek had even convinced the president of General Foods (which had acquired Maxwell House) to use Raggedy Ann in its advertising, so taken was he with Raggedy Ann's homespun image. Few if any print ads, however, actually incorporated Gruelle artwork.

By the 1940s and continuing into the 1950s, companies were freely using images of Gruelle's folksy dolls and toys to pitch cereals, crackers, bread, soap, and gelatine desserts, as well as electric refrigerators, washable flooring, easy-care fabrics, and foodstuffs.

Advertisers (some of which secured licenses, others that did not) also discovered that Gruelle's gentle, guileless characters could take the edge off hard-to-glamorize products and services such as oil drilling,

(249)

Good Housekeeping, *March 1931*.

(250)

Fortune, *April 1941. Artwork by Worth Gruelle.*

moving-van lines, food containers, and chemical research. It was not surprising that most of the advertising incorporating the Raggedys appeared in *Life, Saturday Evening Post,* and *Ladies' Home Journal*—popular magazines celebrating a wholesome American way of life.

(251)

Every mother knows the sparkling goodness of Canada Dry Ginger Ale—now surprise the family with a variety of Canada Dry's 10 tasty flavors.

CANADA DRY

Ladies' Home Journal, *1954.*

(252)

During the 1950s, artist Dick Sargent incorporated Raggedys in many ads for Post cereal.

It is fitting that Frank Luther should collaborate here with the late Johnny Gruelle—the man whose imagination brought Raggedy Ann to life—for Frank and Johnny have much in common as regards the business of making children happy.

Budd Fielding White
Raggedy Ann's Sunny Songs
1946

There were also changes afoot for the Raggedy Ann and Andy dolls. Though by all accounts the Gruelle family and Georgene Novelties had enjoyed a congenial two-decade-long relationship, in order to take advantage of the rapidly expanding playthings marketplace the Gruelles decided to place the production and marketing of the Raggedy dolls in the hands of a larger company.

They turned to the Knickerbocker Toy Company of Middlesex, New Jersey, which, since 1960, had been producing authorized Raggedy Ann and Raggedy Andy non-doll merchandise, including toy puppets, "stuff-and-lace" sewing kits, plastic dolls, Christmas pillows, pajama bags, hassocks, and marionettes. By June of 1963, Knickerbocker became the sole manufacturer authorized to produce Raggedy Ann and Andy stuffed rag dolls.

The transitions to Bobbs-Merrill and Knickerbocker were significant turning points in the merchandising history of Raggedy Ann and Andy, marking the end of an era directly connected to Johnny Gruelle. The transitions inaugurated the beginning for Gruelle's characters of a new, more broadly based commercial life, one driven by expanded product lines, multi-venue marketing, overseas manufacturing, and myriad re-interpretations of the Raggedys' designs.

Raggedy-related merchandising would eventually spread into a diaspora of far-flung interests, activities, and products of varying quality. The appeal and relevance of Gruelle's characters and creations would be alternately lauded and dismissed. Their tender, nostalgic qualities at times would be upheld and celebrated; at other times, they would be overlooked for not being cutting-edge enough to appeal to new generations and therefore not worthy of licensing attention.

(253)

After an unauthorized use of Raggedy Ann in a television ad for Alka-Seltzer, Miles Laboratories agreed to pay $3,000 to the Johnny Gruelle Company for a one-year license to produce advertising such as this, which appeared in the December 6, 1958, issue of Saturday Evening Post.

It is my belief that The Johnny Gruelle Co. proceed vigorously against any activities which they think are an infraction of their rights.

William Erskine to A. G. Saalfield
1940s

(254)

Banner Plastics manufactured boxed dish sets (metal round dishes and square trays) and tea sets (plastic teapot, metal cups, saucers, and utensils). Most sets bore the date 1959. (Candy Brainard)

(255)

(256)

(257)

Jolly Jan and Happy Hank
Know the Raggedy Way.
Everybody wants them 'round
When it's time to play.

"The Raggedy Way"
Raggedy Ann + Andy comic book
1954

Collegeville Flag and Manufacturing Company offered several lines of Raggedy masquerade costumes. Shown here are the insignia on a 1940s Raggedy Andy costume (top left); 1940s formed gauze mask (top right); and 1950s celluloid mask and cloth costume in box (bottom). (Author's Collection/Candy Brainard)

(258)

(259)

(260)

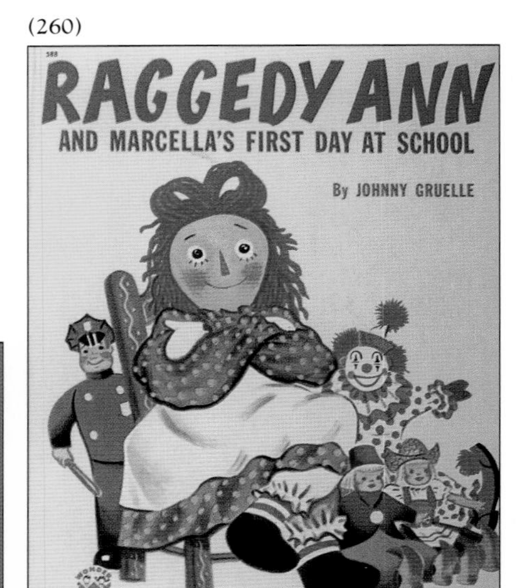

(261)

(262)

The Wonder Book adaptations of the Raggedy stories were also issued as foreign-language editions (translated into German, Swedish, and Danish) and in spiral-bound "Sculptured Cover" editions.

Surges in Raggedy merchandising were met with enthusiasm by consumers and collectors eager to recapture a part of their childhood. Lulls in licensed product production most often coincided, as one might expect, with a spate of unauthorized merchandise. It was a fast-paced, unpredictable, but not altogether unfamiliar era for the legacy Johnny Gruelle had spawned.

Through cycles of inattention and revival, Gruelle's characters have held fast to their unique niche in an ever-evolving playthings marketplace—a niche begun on a wintry morning nearly one hundred years ago when a young cartoonist strode into the *New York Herald,* convinced he could strike the fancy and capture the hearts of a modern audience with his special blend of the whimsical and the old-fashioned.

(263)

"George" and "Georgene" were Georgene Novelties' reaction to having lost their license to produce authorized Raggedy Ann and Andy dolls. (Strong Museum)

A special envelope prepared by Gruelle for his old Hoosier buddy Ed Soltau. (Jessie Soltau Corbin)

Commercial and Publishing Licenses Granted by the Johnny Gruelle Company

The following is an alphabetical listing of all known licenses granted by the Johnny Gruelle Company. Dates refer to license duration, although contract period is not necessarily indicative of product manufacturing or availability dates. Often production ceased before a license expired or items were retailed after licenses had expired.

Artograph Company/Rapid Cutting Company
Brooklyn, NY
(June 1, 1942-July 31, 1945)
Framed pictures priced at retail from 10 cents to $1.

Banner Brothers, Inc.
New York, NY
(May 27, 1941-May 7, 1943)
Children's handbags for sale and distribution through chain stores only, at retail price of 10 and 25 cents.

Banner Plastics Corporation <>**
Patterson, NJ
(Contract #1: November 13, 1958-December 31, 1960) Packaged toy tea sets (plastic and metal).
(Contract #2: February 5, 1960-December 31, 1961) Packaged toy appliance sets (plastic and metal).

Barnes Advertising Agency, Inc.
Milwaukee, WI
(March 1, 1945-March 1, 1946)
Billboard advertisements and store displays in Racine and Milwaukee, WI for John Graf Company ("Gran'pa Graf's Creamy Top Root Beer").

Barr Rubber Products Company
Sandusky, OH
(March 26, 1941-December 1, 1953)
Rubber toys including balloons, tumbaloons, inflated toys, jack sets, and balls.

Beacon Manufacturing Company
Swannanoa, NC
(May 5, 1941-July 1, 1945)
Cotton and wool baby and doll blankets, to be sold exclusively through Montgomery Ward.

Book-of-the-Month Club
New York, NY
(August 26, 1954-January 1, 1955)
One greeting card (no. 14) in Christmas card series. Paid one-time $500 fee.

Braka & Company
New York, NY
(November 20, 1945-December 31, 1948)
Children's handkerchiefs to retail for 10 and 25 cents.

Chicago Printed String Company
Chicago, IL
(March 12, 1953)
Infringement use of Raggedy Ann design on wrapping paper. Paid one-time fee of $250.

Citroen Art Gallery
New York, NY
(August 18, 1945-December 31, 1946)
Bean bags, non-metallic waste baskets, children's bibs, place mats, trays, coat hangers, luminous unframed wall decorations.

Cyrus Clark Company <>**
New York, NY
(March 17, 1958, automatically renewed)
Printed cotton drapery fabric only, not to include other fabric types commonly known as percales, flannelettes, plisses, lawn, and dimity.

Collegeville Flag and Manufacturing Company <>**
Collegeville, PA
(December 12, 1940-August 19, 1960)
Masquerade costumes and play suits for children and adults; "character" cloth to be used exclusively with the sale of merchandise.

Crooksville China Company
Crooksville, OH
(December 2, 1941-December 31, 1952)
Juvenile china tableware.

Decca Records, Inc. <>**
New York, NY
(March 23, 1944, automatically renewed)
Three double-faced phonograph records, with stories from *Raggedy Ann in the Magic Book, Raggedy Ann and the Golden Butterfly,* and *Raggedy Ann and Andy and the Nice Fat Policeman.*

Nelson Doubleday, Inc. <>**
New York, NY
(Contract #1: July 26, 1957)
Permission to excerpt *Raggedy Andy Stories,* $250.
(Contract #2: September 30, 1958)
Permission to excerpt *Raggedy Andy Stories,* $500.

M. A. Donohue & Company <*> <>**
Chicago, IL
(1939-August 19, 1960)
File agreements for publishing, including reprinting of "Sunny Book" titles, *Johnny Gruelle's Golden Book,* and selected Raggedy Ann series titles.

Duenewald Printing Corporation
New York, NY
(August 3, 1943-December 31, 1946)
Animated book *Raggedy Ann and Raggedy Andy* to be sold only to Saalfield Publishing Company.

Mrs. John S. Elliott
Miami Shores, FL
(May 11, 1953, expires on 30 days notice)
Ceramic figurines.

The Firth Carpet Company
New York, NY
(July 7, 1941-August 15, 1944)
Trademark 391,950 registered December 2, 1941 (serial #444,331)
Carpets, rugs, and "rag" rugs.

Fisher-Price Toys, Inc.
East Aurora, NY
(January 30, 1941-March 1, 1945)
Push-and-pull toys.

Fleischer Studios, Inc. *(See Paramount Pictures, Inc.)*

F. A. Foster & Company
Boston, MA
(January 22, 1942-December 31, 1951)
Printed cotton drapery fabrics, in "Puritan Prints" line. License excludes printed fabrics known as percales, plisses, flannelettes, lawn, and dimity.

Gardner Displays Company
Pittsburgh, PA
(December 12, 1940-December 31, 1949)
Displays of paper mache, rubber (materials other than lithographed, printed, or other processes applied to paper or cardboard).

Georgene Novelties, Inc. <*> <>**
New York, NY
(Contract #1: November 23, 1938-March 18, 1941)
Stuffed rag dolls and doll clothing (Raggedy Ann, Raggedy Andy and Beloved Belindy).
(Contract #2: March 18, 1941-December 31, 1954) Stuffed rag dolls (Beloved Belindy and Camel with the Wrinkled Knees).
(Contract #3: February 28, 1948-December 31, 1950) Plastic toy handbags and handbags.

Glendser Textile Corporation
New York, NY
(November 12, 1949-December 31, 1950)
Scarves using Raggedy Ann and Andy characters.

Grosset & Dunlap
New York, NY
(April 10, 1952-June 18, 1955)
"Notch-Ems" and "Doodle Boxes," copyright June 17, 1955.

Halsam Products Company, Inc.
Chicago, IL
(November 25, 1940-December 31, 1952)
Children's wooden embossed "Safety Blocks" in two sizes.

William H. Hirsch Manufacturing Company
Hollywood, CA
(April 7, 1943-December 31, 1949)
Glazed ceramic and plaster of Paris figurines, bookends, planters, salt and pepper shakers, paper weights, and bells.

Holgate Brothers Company
Kane, PA
(April 18, 1941-February 28, 1946)
Wooden "Rocky Toys" and "Concentration Toys."

Inez Holland House
New York, NY
(August 31, 1954-December 31, 1955)
Children's bibs, pinafores, robes, and creepers.

Hutchinson & Company (Publishers), Ltd.
London, England
(March 13, 1941-June 20, 1952)
Reprints of selected Raggedy Ann books already published in U.S.A.

Kagran Corporation
New York, NY
(January 11, 1952)
Assignment of copyright for use of likeness of Raggedy Ann on phonograph record jacket for Howdy Doody record, *Howdy Doody Dos and Don'ts* (RCA Victor #Y445).

continued on next page

R. W. Kellogg Company
Three Rivers, MI
(September 6, 1950-August 31, 1955)
Use of Raggedy Ann name on flowering plants (azaleamums) to sell for 35 cents each.

Keeshan-Miller Enterprises Corporation
New York, NY
(March 19, 1956-March 19, 1957)
Television and radio broadcast rights.

Kits, Inc.
Miamisburg, OH
(May 24, 1941-December 31, 1950)
Raggedy Ann and Andy "Sticker Kits."

Richard G. Krueger, Inc.
New York, NY
(October 25, 1940-December 31, 1950)
Baby, autograph, and snapshot albums; nursery plaques, lamps, bookshelves, laundry bags, and dress hangers.

Linen Embroidery Company
(Edward D. Brown)
New York, NY
(July 22, 1947-December 31, 1951)
Appliqued towel sets, crib sheets, and pillow cases.

Madmar Quality Company <*>
Utica, NY
(November 15, 1954-December 21, 1955)
Permission to use Volland Company endpapers.

Magic Mirror Movies Company <>**
New Canaan, CT
(June 20, 1957-December 31, 1959)
Toy Red Raven movie record.

McCall Corporation <>**
New York, NY
(July 1, 1940, automatically renewed until notice given by either party)
Printed or stamped home sewing patterns.

McLoughlin Bros., Inc.
Springfield, MA
(June 21, 1940-December 31, 1947)
Children's paint, coloring, cut-out, tracing, novelty, and story/picture books for syndicate trade.

Meldrum & Fewsmith, Inc.
Cleveland, OH
(September 25, 1940-December 31, 1941)
Permission to use Raggedy Ann in full-page ad in *Fortune* (April 1941 issue) and in advertising folder for Gilbert Paper Company.

Miles Laboratories, Inc.
Elkhart, IN
(December 11, 1958-January 2, 1959)
TV and other advertising used in selling Alka-Seltzer.

Milton Bradley Company <>**
Springfield, MA
(June 21, 1940, renewed automatically each year, for one year, provided licensee pays minimum $1,000 in license fees)
Boxed paint and crayon sets, picture puzzles, paper dolls, bubble sets, and games.

Ruth Alexander Nichols
Westfield, NJ
(September 28, 1950-July 1, 1952)
Photographs of characters for advertising baby pants for International Latex Corporation of New York. Paid one-time fee of $750.

Old King Cole Displays
Louisville, OH
(February 1, 1952-February 1, 1953)
Displays of paper maché, rubber, etc. not printed on paper or cardboard.

Owens-Illinois Glass Company
Toledo, OH
(June 20, 1941-December 31, 1952)
Glass containers, including tumblers and metal containers for packing purposes.

Paramount Pictures, Inc.
New York, NY
(Contract #1: April 18, 1940). With Fleischer Bros. Studios, Inc., Miami, FL, for full motion picture rights.
(Contract #2: December 2, 1943). Transfer of Fleischer contract.

R.K.O. Radio Pictures, Inc.
New York, NY
(February 21, 1945)
Permission to use song "Raggedy Ann" in motion picture "Heavenly Days." Paid one-time fee of $750.

Radio Corporation of America
(R.C.A. Victor Division)
Camden, NJ
(Contract #1: December 16, 1947-December 31, 1948, with option to renew)
Phonograph records (*Songs of Raggedy Ann*) of songs and music taken from *Raggedy Ann's Sunny Songs* and *Raggedy Ann's Joyful Songs* folios.
(Contract #2: November 3, 1947-December 31, 1948, with option to renew)
Radio shows promoting R.C.A. children's records.

Raggedy Ann Corporation
Chicago, IL
(June 20, 1941)
Agreement to clarify rights in field of foods.

C. A. Reed Company
Williamsport, PA
(Contract #1: February 3, 1941-December 31, 1957) For party favors, nut cups, and snapping mottoes.
(Contract #2: November 1, 1956-August 19, 1960) For paper party table ensembles (napkins, plates, etc.)

Saalfield Publishing Company <>**
Akron, OH
(Contract #1: January 1, 1944-December 31, 1957) Limp cover coloring, cut-out, paper doll, and "shaped" paint books to sell for not more than 10 cents.
(Contract #2: December 20, 1944-January 1, 1951) Story and picture books to sell for not more than 29 cents.

The Sterling Company
Chicago, IL
(April 25, 1942-June 30, 1942)
Cotton dresses in misses' and women's sizes 12-20.

Thomas Textile Company
(1941-)
Children's bedspreads, sweatshirts, and polo shirts.

Tilton & Cook Company
Leominster, MA
(May 2, 1941-May 31, 1942)
Children's plastic jewelry pins, for sale to syndicate trade, notion jobbers, and mail order houses only.

Trojan Maid Company (Phillip A. Appelbaum)
Troy, NY
(September 26, 1941-December 31, 1950)
Cotton or rayon girls' dresses, brother-and-sister suits, sun and play suits, sizes 1-12.

U.S. Fiber & Plastic Corporation
Stirling, NJ
(September 8, 1948-December 31, 1950)
Inflatable vinyl characters, wading pools, balls, swimming rings.

P. F. Volland Company <*>
Joliet, IL
(November 1, 1940-February 28, 1949)
Children's birthday and holiday greeting cards.

Western Printing & Lithographing Company <>**
Racine, WI
(Contract #1: March 2, 1942-March 27, 1946) Raggedy Ann comic books and magazines.
(Contract #2: August 31, 1942-August 31, 1946) Mr. Twee Deedle
(Contract #3: October 25, 1945-December 31, 1957) One-shot Raggedy Ann comic books.

White & Wyckoff Manufacturing Company
Holyoke, MA
(November 13, 1940-December 31, 1943)
Children's writing paper and boxes, and greeting cards.

Whitney Manufacturing Company
New York, NY
(September 26, 1941, terminates on 60 days written notice)
Hang tickets to be sold to Raggedy Ann licensees in apparel field.

Wonder Books, Inc. (Grosset & Dunlap, Inc.) <>**
New York, NY
(Contract #1: June 25, 1951-October 15, 1960) Picture books to sell for not less than 25 cents nor more than 39 cents.
(Contract #2: March 3, 1960, for three years after first publication) "Read-Aloud" book to sell for 39 cents per copy.

Lawrence Wynn
New York, NY
(July 11, 1950-July 30, 1955)
Preparation of material and scripts for television broadcast.

<*> an earlier license or authorization was granted by Johnny or Myrtle Gruelle and/or the P. F. Volland Company.

<**> license was renewed/reactivated by the Bobbs-Merrill Company on August 19, 1960.

Myrtle, Worth, and Richard Gruelle. (Miami Herald)

Principal Sources

Articles and Chapters

Burdick, Loraine. "Johnny Gruelle and the Raggedys." In *The Antique Trader Weekly's Book of Collectible Dolls*, ed. by Kyle D. Husfloen. Dubuque, IA: The Babka Publishing Company, 1976.

Johnson, Hope. "A Bright New World for Raggedy Ann." *New York World Telegram and Sun Feature Magazine* (February 18, 1961).

Langer, Mark. "Working at the Fleischer Studio: An Annotated Interview with Myron Waldman." *The Velvet Light Trap*, no. 24 (Fall 1989).

Williams, Martin. "Some Remarks on Raggedy Ann and Johnny Gruelle." *Children's Literature 3*, ed. by Francelia Butler. Philadelphia: Temple University, 1974.

Zipes, Jack. "Johnny Gruelle." In *Dictionary of Literary Biography, Volume 22: American Writers for Children, 1900-1960*. Ed. by John Cech. Detroit: Gale Research, 1983.

Books, Booklets, Exhibition Catalogues, Pamphlets, and Videotapes

Adams, Margaret, ed. *Collectible Dolls and Accessories of the Twenties and Thirties from Sears, Roebuck and Company Catalogs*. New York: Dover Publications, 1986.

Cieslik, Jurgen and Marianne. *Button in Ear: The History of the Teddy Bear and His Friends*. Julich, West Germany: Marianne Cieslik Verlag, 1989.

Crafton, Donald. *Before Mickey: The Animated Film 1898-1928*. Cambridge and London: MIT Press, 1982.

An Exhibit of Indiana Art: Contemporary & Retrospective. Catalogue for art exhibition at Tomlinson Hall, Indianapolis, IN, April 27-May 9, 1903.

Formenek-Brunell, Miriam. *Made to Play House: Dolls and the Commercialization of American Girlhood, 1830-1930*. New Haven and London: Yale University Press, 1993.

Garrison, Susan Ann. *The Raggedy Ann and Andy Family Album*. West Chester, PA: Schiffer Publishing, 1989.

Hall, Patricia. *Johnny Gruelle, Creator of Raggedy Ann and Andy*. Gretna, LA: Pelican Publishing Company, 1993.

Hall, Patricia <writer> and Leonard Swann/Gayle O'Neal <producers/directors>. *Raggedy Ann and Andy: Johnny Gruelle's Dolls with Heart* <90-minute videotape>. Norfolk, VA; Sirocco Productions, Inc.: 1995.

Murray, John J. and Bruce R. Fox. *Fisher-Price 1931-1963: A Historical, Rarity, Value Guide (Second Edition)*. Florence, AL: Books Americana, 1991.

O'Brien, Richard. *The Story of American Toys*. New York: Abbeville Press, 1990.

Solomon, Charles. *The History of Animation: Enchanted Drawings*. New York: Random House, 1994.

Swann, Leonard <writer> and Leonard Swann and Gayle O'Neal <producers/directors>. *The Dollmakers: Women Entrepreneurs 1865-1945* <90-minute videotape>. Norfolk, VA: Sirocco Productions, Inc., 1995.

Tabbat, Andrew. *The Collector's World of Raggedy Ann and Andy—Volume I*. Annapolis, MD: Gold Horse Publishing, 1996.

Tabbat, Andrew. *The Collector's World of Raggedy Ann and Andy—Volume II*. Annapolis, MD: Gold Horse Publishing, 1997.

Williams, Martin. *Hidden in Plain Sight: An Examination of the American Arts*. New York and Oxford: Oxford University Press, 1992.

Young, Mary. *A Collector's Guide to Paper Dolls: Saalfield, Lowe, Merrill*. Paducah, KY: Collector Books, 1980.

Zipes, Jack, trans., ed. *The Complete Fairy Tales of the Brothers Grimm*. New York: Bantam Books, 1987.

General Reference Works

Bachs, Jean. *The Main Street Dictionary of Doll Marks*. Pittstown, NJ: The Main Street Press, 1985.

Coleman, Dorothy S., Elizabeth A. Coleman, and Evelyn J. Coleman. *The Collector's Encyclopedia of Dolls—Volume 1*. New York: Crown, 1968.

Coleman, Dorothy S., Elizabeth A. Coleman, and Evelyn J. Coleman. *The Collector's Encyclopedia of Dolls—Volume 2*. New York: Crown, 1986.

Commire, Anne. *Something About the Author. Facts and Pictures about Authors and Illustrators of Books for Young People*, vol. 35. Detroit: Gale Research, 1984.

Cunningham, Jo. *The Collector's Encyclopedia of American Dinnerware*. Paducah, KY: Collector Books, 1982.

Hart, Luella. *Directory of United States Doll Trademarks, 1888-1968*. Self-published, 1968.

Lenburg, Jeff. *The Encyclopedia of Animated Cartoon Series*. Westport, CT: Arlington House, 1981.

Mott, Frank Luther. *A History of American Magazines*. Vols. 3, 4, and 5. Cambridge: Belknap Press of Harvard University Press, 1957-68.

The National Union Catalogue—Pre-1956 Imprints. Mansell, 1972.

Official Gazette, U.S. Patent Office. Washington, DC: U.S. Government Printing Office, 1910-60.

Overstreet, Robert M. *The Overstreet Comic Book Price Guide*. 25th Anniversary Edition (25th ed.). New York: Avon Books, 1995.

Tebbel, John. *A History of Book Publishing in the United States*. Vols. 2, 3, and 4. New York and London: R. R. Bowker, 1975-81.

Magazines, Journals, and Newsletters

The American Magazine, College Humor, Cosmopolitan, The Delineator, Designer, Good Housekeeping, Growing Point, Illustrated Sunday Magazine, Fortune, John Martin's Book, Judge, Ladies Home Journal, The Ladies' World, Leslie's Illustrated Weekly, Life (the humor weekly), *Little Folks, McCall's, McCall's Fashion News, McCall's Needlework, The Moving Picture World, Physical Culture, Pictorial Review, Playthings, "Rags," Publishers Trade List Annual (PTLA), Publisher's Weekly, Schoenhut Newsletter, Teddy & Toy Review, Toys & Novelties, Woman's World.*

Newspapers

The Boston Globe, The Boston Post, The (Wilmington, DE) *Bulletin, The Chicago Tribune, The Cleveland Press, The* (Indianapolis) *Daily Sentinel, The* (Ashland, OR) *Daily Tidings, The* (Indianapolis) *Journal, The Indianapolis News, The Indianapolis Star, The Indianapolis Sun, The Joliet* (IL) *Herald-Times, The Los Angeles Herald, The Louisville Courier-Journal, The Miami Daily News, The Miami Herald, The New Canaan* (CT) *Advertiser, The New Canaan* (CT) *Messenger, The New York Herald, The New York Herald-Tribune, The New York Times, The News of Delaware County* (PA), *The Norwalk Hour, The Oakland Tribune, The Paris Herald, The* (Memphis) *Commercial Appeal, The* (Indianapolis) *People, The* (Rochester, NY) *Times-Union, The Tennessean, The Washington Post, The Washington* (DC) *Star, The Wilton* (CT) *Bulletin.*

Company and Mail-Order Catalogues

Decca Records, M. A. Donohue & Company, Fisher-Price Company, Gimbel Brothers, E. I. Horsman Company, Madmar Quality Company, Marshall-Field Company, McCall Pattern Company, Milton Bradley Company, R.C.A. Victor Records, Sears-Roebuck Company, Steiff, P. F. Volland Company.

Oral Interviews

Vital to the author's research were in-person and telephone interviews, conducted with the following individuals between 1987 and 1996: Bill Blackbeard, Candy Brainard, Jane Marcella Ferguson Braunstein, Lynda Oeder Britt, Charles Carton, John Cech, Jane Gruelle Comerford, Jessie Soltau Corbin, Monica Borglum Davies, Betty MacKeever Dow, Kathy and John Ellis, Flora Faraci, Marge Livingston Grinton, Kim Gruelle, Kit Gruelle, Ruth Gruelle, Worth and Suzanne Gruelle, Kathy Erskine Jenkins, Anne laDue Hartman Kerr, Barbara Lauver, Janet Smith Leach, Richard Marschall, Jacki Payne, Caroline Cox Scheffey, Peggy Slone, Patricia Smith, Andrew Tabbat, Gloria Timmell, Joni Gruelle Wannamaker, Martin Williams.

Key Unpublished Materials

Rounding out the research database were many unpublished resources, which included:

Correspondence: Letters and memos generated by James Gordon Bennett, Lynda Oeder Britt, David Carton, David Laurence Chambers, Jessie Soltau Corbin, Howard Cox, Charles Dillingham, Ethel Fairmont, Mollye Goldman, Johnny Gruelle, Justin Gruelle, Myrtle Gruelle, Suzanne Gruelle, Worth Gruelle, H. H. Howland, Ray Powell, John W. Thompson, Gordon Volland, Martin Williams, Fred Wish, Jack Zipes.

Memoirs, Diaries, Family Histories: Justin C. Gruelle (family history, n.d.), Kim Gruelle ("The Gruelle Tradition: Four Generations of Artists", n.d.), Myrtle Gruelle (diary, 1923; "I Am But One," ca. 1946; lecture notes, 1951), Richard Buckner Gruelle (autobiographical letter, 1914).

Interview notes and Transcriptions: John Randolph Bray (Mark Langer, interviewer, 1975), Clifton Meek (Martin Williams, interviewer, 1969).

Bibliographic Checklists: Ray Powell (ca. 1968), Martin Williams (n.d.).

Other Sources

In addition to the aforementioned sources, the author researched selected company records of the Bobbs-Merrill, P. F. Volland, M. A. Donohue, and Johnny Gruelle companies; patent, trademark, and copyright records held by the U.S. Copyright Office; selected proceedings of the U.S. circuit, district, and supreme courts; as well as vital records and other publicly and privately held records pertaining to the Gruelle family and Johnny Gruelle's published and licensed works.

Index

Price Guide

This price guide is designed to be used with *Raggedy Ann and More: Johnny Gruelle's Dolls and Merchandise*. Entries are keyed to correspond numerically with selected illustrations in the book. The suggested price values are **averages,** computed from numerous sources, including but not limited to individual collectors, collectible and antique dealers; collector conventions and gatherings; and mail order and online sale lists and auctions. Whenever possible, calculations are based on actual **sale** (rather than **asking**) prices, and care has been taken to prevent extraordinarily low or high prices from skewing the averages.

Collectible values are influenced by age, scarcity, original manufacturing numbers, provenance, collector demand, media coverage, sentimental value, geographic region, and general strength of the collectibles marketplace. With these factors in mind, the ultimate determinant of value is an item's condition. The values in this price guide are presented in ranges—the lesser dollar amount applies to items in "fair" condition; the greater, to items in "excellent-to-fine" condition.

This price guide is based on independent research by Patricia Hall. Its compilation and publication were not underwritten or sponsored by any publisher, manufacturer, or trademark holder for Raggedy Ann and Andy. These suggested values are guidelines and are not intended to set prices for commerce. Neither the author nor Pelican Publishing Company assumes responsibility for losses that may occur as a result of relying on the information that follows.

(1) Doll . . . . $1,400-$6,500
(2) Dolls, each . . . . $800-$3,000
(3) Booklet . . . . $60-$125
(4) Comic page . . . . $25-$40
(5) Comic page . . . . $30-$50
(6) (7) (8) Dolls, each . . . . $5,000-$8,000
(9) Book . . . . $125-$300
Book in dust jacket . . . . $175-$350
(10) Book . . . . $75-$100
Book in dust jacket . . . . $125-$150
(11) Book . . . . $40-$80
Book in dust jacket . . . . $90-$130
(12) Book . . . . $90-$140
Book in dust jacket . . . . $140-$190
(13) (14) Books, each . . . . $500-$1,200
(15) Boxed set . . . . $350-$600
(16) Book . . . . $65-$125
Book in dust jacket . . . . $135-$175
(17) (18) (19) Books, each . . . . $50-$75
Book in dust jacket . . . . $100-$125
(20) Sheet music . . . . $65-$125
(21) Fabric swatch . . . . $45-$90
(22) Toy . . . . $100-$250
(23) Toys, each . . . . $400-$800
Toy in box . . . . $650-$1,050
(24) Book . . . . $125-$200
Book in box . . . . $175-$250
(25) Calendar . . . . $50-$75
(26) Booklet . . . . $75-$200
(27) Poster . . . . $150-$300
(28) (29) Books, each . . . . $65-$135
Book in box . . . . $110-$185
(30) Book . . . . $300-$500
Book in box . . . . $450-$650
(31) Booklet . . . . $75-$125
(32) (33) (34) Books, each . . . . $75-$200
Book in box . . . . $150-$225
(35) (36) (37) (38) Flyers, each . . . . $25-$40
(39) Doll . . . . $1,400-$6,500
(40) Book . . . . $75-$150
Book in box . . . . $125-$200
(41) Doll . . . . $1,400-$6,500
Book . . . . $75-$150
(42) (43) (44) (45) Dolls, each . . . . $800-$2,000
(46) (47) (48) (49) Dolls, each . . . . $800-$2,000
(50) Magazine cover . . . . $20-$25
(51) Flyer . . . . $20-$35
(52) (53) (54) (55) (56) Dolls, each . . . . $850-$2,500
(57) (58) Dolls, each . . . . $750-$2,000
(59) (60) Dolls, mint, each $2,700

(61) (62) Dolls, each . . . . $650-$2,000
(63) Dolls, mint, each $2,700
(64) Book . . . . $75-$150
Book in box . . . . $125-$200
(65) Flyer . . . . $25-$40
(66) (67) (68) (69) Greeting cards, each . . . . $30-$65
(70) Doll . . . . $1,200-$2,700
(71) Magazine page . . . . $20-$25
(72) Book . . . . $125-$200
Book in box . . . . $175-$250
(73) Magazine page . . . . $25-$30
(74) Books, each . . . . $65-$135
Book in box . . . . $110-$185
(75) Doll . . . . $850-$2,400
(76) Book . . . . $125-$200
Book in dust jacket . . . . $175-$250
(77) Dolls, each . . . . $650-$2,500
(78) Dolls, each . . . . $650-$2,500
(79) (80) (81) Sheet music, each . . . . $20-$45
(82) Magazine page . . . . $10-$12
(83) Shelf . . . . $850-$1,500
(84) (85) Dolls, each . . . . $900-$3,000
(86) Book . . . . $125-$225
Book in box . . . . $200-$300
(87) Comic page . . . . $15-$20
(88) Magazine page . . . . $12-$18
(89) Boxed set . . . . $250-$450
(90) Sheet music . . . . $70-$120
(91) Doll . . . . $700-$2,000
(92) Doll . . . . $500-$1,500
(93) Doll . . . . $700-$2,000
(94) Doll . . . . $500-$1,500
(95) Doll . . . . $300-$900
(96) Dolls, mint, each . . . . $3,000-$3,500
Bench . . . . $800-$2,000
(97) (98) (99) Dolls, each . . . . $700-$2,000
(100) Book . . . . $80-$125
Book in box . . . . $100-$200
(101) Puzzle . . . . $70-$90
(102) (103) (104) Boxed sets, each . . . . $250-$400
(105) Booklet . . . . $70-$125
(106) Book . . . . $75-$125
Book in dust jacket . . . . $100-$150
(107) Bookends, each . . . . $500-$900
(108) Book . . . . $65-$140
Book in dust jacket . . . . $90-$165
(109) Book . . . . $45-$70
Book in dust jacket . . . . $80-$95
(110) Boxed set . . . . $250-$400
(111) Book . . . . $150-$250
Book in dust jacket . . . . $200-$300
(112) Book . . . . $100-$200
Book in dust jacket . . . . $150-$250
(113) Catalogue . . . . $20-$40
(114) Book . . . . $40-$90
Book in dust jacket . . . . $65-$115
(115) Magazine page . . . . $10-$15
(116) Program . . . . $65-$100
(117) (118) (119) Clippings, each . . . . $7-$10
(120) Book . . . . $50-$100
(121) Book . . . . $40-$100
(122) Book . . . . $75-$135
Book in dust jacket . . . . $125-$185
(123) Cut-outs (2) . . . . $20 -$40
(124) Booklet, uncut . . . . $150-$350
(125) Doll . . . . $2,500-$4,000
(126) Doll, undressed and signed . . . . $1,000-$3,500
(127) Doll, mint $5,000
(128) (129) Dolls, each . . . . $900-$2,000
(130) Doll . . . . $1000-$2,500
(131) (132) Dolls, each . . . . $900-$2,000
(133) Doll with altered logo . . . . $900-$2,000
(134) (135) Mint dolls, each $3,500
(136) (137) (138) (139) Booklets, each . . . . $30-$45
(140) (141) Greeting cards, each . . . . $15-$20
(142) (143) (144) Booklets, each . . . . $30-$45
(145) Towel . . . . $60-$100
(146) Puzzle . . . . $20-$35
(147) (148) (149) Dolls, each . . . . $150-$400
(150) (151) Dolls, each . . . . $400-$800
(152) Dolls, each . . . . $400-$600
(153) Dolls, each . . . . $300-$450
(154) Dolls, each . . . . $500-$800
(155) Dolls, each . . . . $400-$700
(156) Dolls in boxes, each . . . . $200-$350
(157) (158) Dolls, each . . . . $100-$250
(159) Dolls, each . . . . $75-$150
(160) Doll . . . . $200-$300
(161) (162) (163) (164) Books, each . . . . $50-$85
Book in dust jacket . . . . $75-$110
(165) (166) (167) (168) Books, each . . . . $50-$85
Book in dust jacket . . . . $75-$110

(169) Poster .... $75-$180
(170) (171) Sheet music, each .... $60-$100
(172) Photo album .... $45-$125
(173) Toy .... $400-$900
(174) Dolls, each .... $100-$450
(175) Dolls, each .... $50-$300
(176) (177) Patterns, each .... $20-$75
(178) Pattern .... $35-$85
(179) Doll .... $100-$350
(180) Pattern .... $40-$90
(181) Book .... $35-$65
(182) Doll .... $750-$3,000
(183) Boxed set .... $50-$100
(184) Doll .... $600-$2,500
(185) Doll .... $500-$2,000
(186) Greeting card .... $25-$45
(187) (188) (189) (190) Boxed games/sets, each .... $150-$275
(191) Boxed game .... $35-$85
(192) Puzzle .... $20-$40
(193) Dishes, each .... $25-$85
(194) (195) Record albums, each .... $45-$100
(196) Blankets, each .... $60-$125
(197) Record album $60-125
(198) Toy .... $500-$850
(199) Toy set .... $150-$250
Toy set in box .... $250-$350
(200) Handbag .... $70-$150
(201) (202) (203) Framed prints, each .... $20-$45
(204) Planter .... $40-$75
(205) (206) Fabric swatches, each .... $25-$75
(207) Booklet .... $40-$100
(208) Booklet .... $65-$150
(209) (210) (211) Books, each .... $45-$70
Book in dust jacket .... $60-$85
(212) (213) Books, each .... $35-$65
(214) Boxed set .... $120-$350
(215) (216) (217) Records, each .... $40-$85
(218) (219) Handkerchiefs, each .... $70-$125
(220) (221) Greeting cards, each .... $20-$40
(222) (223) Greeting cards, each .... $30-$50
(224) Book .... $65-$170
Book in dust jacket .... $95-$200
(225) Booklet .... $30-$90
(226) (227) Books, each .... $40-$60
Book in dust jacket .... $70-$85
(228) (229) Books, each .... $40-$60
Book in dust jacket .... $70-$85
(230) (231) Booklets, each .... $15-$35
(232) (233) Boxed sets, each .... $90-$150
(234) Booklet .... $25-$45
(235) Booklet .... $20-$40
(236) Booklet, uncut .... $40-$65
(237) Record .... $30-$70
(238) Booklet .... $20-$30
(239) (240) (241) (242) Booklets, each .... $15-$25
(243) Boxed set .... $300-$500
(244) Ball .... $45-$65
(245) Blocks, each .... $10-$20
(246) Boxed set .... $200-$600
(247) Record player in box .... $50-$125
(248) Record .... $15-$30
(249) (250) Magazine pages, each .... $15-$20
(251) (252) Magazine pages, each .... $10-$15
(253) Magazine page .... $10-$15
(254) Boxed set .... $100-$150
(255) Costume .... $25-$50
(256) Mask .... $20-$40
(257) Costume and mask in box .... $50-$100
(258) (259) (260) Books, each .... $15-$20
(261) Book .... $35-$45
(262) Book .... $20-$30
(263) Dolls, each .... $75-$125
In box .... $100-$150